WITHDRAWN

FV

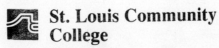

St. Louis Community College

Forest Park
Florissant Valley
Meramec

Instructional Resources
St. Louis, Missouri

WIND ENGINEERING

A Handbook for Structural Engineers

Henry Liu
University of Missouri–Columbia

Prentice Hall, Englewood Cliffs, New Jersey 07632

Library of Congress Cataloging-in-Publication Data

Liu, Henry.
 Wind engineering : a handbook for structural engineers / Henry
Liu.
   p. cm.
 Includes bibliographical references.
 ISBN 0-13-960279-8
 1. Wind-pressure. 2. Buildings--Aerodynamics. 3. Structural
engineering.   I. Title.
TA654.5.L58 1990
624.1'75--dc20   89-48279
                                                             CIP

Editorial/production supervision
 and interior design: *Brendan M. Stewart*
Cover design: *Lundgren Graphics*
Manufacturing buyer: *Kelly Behr*

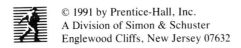

© 1991 by Prentice-Hall, Inc.
A Division of Simon & Schuster
Englewood Cliffs, New Jersey 07632

All rights reserved. No part of this book may be
reproduced, in any form or by any means,
without permission in writing from the publisher.

Printed in the United States of America
10 9 8 7 6 5 4 3 2 1

ISBN 0-13-960279-8

Prentice-Hall International (UK) Limited, *London*
Prentice-Hall of Australia Pty. Limited, *Sydney*
Prentice-Hall Canada Inc., *Toronto*
Prentice-Hall Hispanoamericana, S.A., *Mexico*
Prentice-Hall of India Private Limited, *New Delhi*
Prentice-Hall of Japan, Inc., *Tokyo*
Simon & Schuster Asia Pte. Ltd., *Singapore*
Editora Prentice-Hall do Brasil, Ltda., *Rio de Janeiro*

Contents

Foreword   ix

Preface   xi

1   High Winds and Severe Storms   1

    1.1   Introduction, 1
    1.2   Types of High Winds, 1
    Hurricanes, Typhoons and Cyclones, 1
    Tornadoes, 10
    Thunderstorm Wind, Straight-Line Wind and
        Downbursts, 20
    References, 24

2   Extreme Wind Probability   25

    2.1   Introduction, 25
    2.2   Exceedance Probability and Return Period, 26
    2.3   Probability Density and Distribution Functions, 27
    2.4   Probability of Ordinary Winds, 28
    2.5   Probability of Extreme Winds, 29
    Thunderstorm Wind Probability, 30

Hurricane Wind Probability, 32
Tornado Wind Probability, 34
2.6 Significance of Return Period and Exceedance Probability, 37
References, 39

3 Wind Characteristics 41

3.1 Variation of Wind Velocity With Height and Roughness, 41
Variation of Wind Speed with Height, 41
Variation of Wind Speed with Surface Roughness, 43
Variation of Wind Direction with Height, 46
3.2 Turbulent Characteristic of Wind, 46
Averaging Time, 46
Fastest-Mile Wind, 47
Gust Factors, 48
Turbulence Intensity, 49
Spectrum of Turbulence, 51
Correlation Coefficient of Turbulence, 53
Integral Scales of Turbulence, 54
3.3 Modification of Wind by Topography, Woods, and Structures, 55
Topographical Effect on Wind, 55
Effect of Woods on Wind, 62
Effect of Buildings/Structures on Wind, 62
3.4 Directional Preference of High Winds, 66
References, 67

4 Wind Pressure and Forces on Buildings and Other Structures 69

4.1 Introduction to Wind Pressure, 69
Definition and Units, 69
External and Internal Pressures, 70
Ambient Pressure, 70
Stagnation Pressure, 71
Dimensionless Pressure (Pressure Coefficient), 72
4.2 Wind Pressure on Rectangular Buildings, 76
External Flow and Pressure, 77
Pressure Inside Buildings: The Internal Pressure, 84
Correlation of Pressure Fluctuations, 91
4.3 Wind Pressure on Other Structures, 92

Contents   v

      Circular Cylinders, 92
      Hyperbolic Cooling Tower, 96
   4.4 Wind Forces and Moments, 98
      Cladding Forces, 98
      Drag Force, 98
      Lift Force, 101
      Normal Forces, 103
      Vertical Overturning Moment, 103
      Horizontal Twisting Moment (Torsion), 104
      Shear (Fluid Friction) on Structures, 104
   References, 106

5 Dynamic Response of Structures: Vibration and Fatigue   109

   5.1 Introduction, 109
      Wind Sensitive Structures, 109
      Factors Contributing to Wind-Induced Vibration, 109
      Aerodynamic and Aeroelastic Instabilities, 110
   5.2 Types of Structural Vibration, 111
      Vortex-Shedding Vibration, 111
      Vibration at Critical Reynolds Number, 115
      Across-Wind Galloping, 115
      Wake Galloping, 119
      Torsional Divergence, 120
      Flutter, 122
      Buffeting Vibration, 123
      Wind-Induced Building Vibration, 124
   5.3 Means to Reduce Wind-Induced Vibration, 124
   5.4 Predicting Dynamic Response of Structures, 126
      General Approaches, 126
      Vibration of Cantilever Beam, 129
      Along-Wind Response of Structures Due to Buffeting, 131
      National Frequency and Damping Ratio of Structures, 134
      Across-Wind Response of Structures, 136
      Vibration of Stacks, Chimneys and Towers, 138
   5.5 Wind-Induced Fatigue, 140
      Introduction, 140
      Wind-Induced Fatigue, 141
      Case Study, 142
      Mitigation of Wind-Induced Fatigue, 144
   References, 144

6 Wind Tunnel Tests   147

6.1 Introduction, 147
6.2 Circumstances for Conducting Wind Tunnel Tests, 148
   Cost of Structures, 148
   Likelihood of Wind Problems, 149
   Complex Structures, 149
   Importance of Structures, 149
   Performance Criteria for Structures, 150
6.3 Taxonomy of Wind Tunnels, 151
   Flow Circuit, 151
   Throat Condition, 153
   Pressure Condition, 154
   Wind Speed, 155
   Velocity Profile, 155
   Temperature Stratification, 156
   Cross-Sectional Geometry, 156
   Drive System, 156
   Turbulence Level, 156
   Special purpose Tunnels, 158
6.4 Boundary-Layer Wind Tunnel, 159
6.5 Major Components of Wind Tunnels, 161
6.6 Similarity Parameters and Wind Tunnel Tests, 162
   Similarity of Approaching Flows, 163
   Other Similarity Parameters for Stationary Structures, 163
   Scaling Laws, 165
   Modeling Vibrating Structures, 167
   Gravity Effect on Vibration, 169
   Modeling Internal Pressure, 169
6.7 Wind Tunnel Instrumentation, 170
   Pitot Tube, 170
   Hot-Wire Anemometer, 170
   Manometers, 171
   Pressure Transducers, 171
   Other Sensors, 172
   Data Acquisition Systems, 172
References, 172

7 Building Codes and Standards   173

7.1 Introduction, 173
7.2 ANSI Standard, 175

Brief History, 175
General Approach, 175
Procedure for Wind Load Determination, 176
Comments on ANSI A58.1, 179
7.3 Model Codes, 179
Uniform Building Code (UBC), 180
Basic/National Building Code (BOCA), 180
Standard Building Code (Southern Code), 180
7.4 International Standards, 181
Canadian Standard, 181
British Standard, 182
Australian Standard, 183
Japanese Standard, 185
ISO Standard, 186
European Community Standard, 186
7.5 Prescriptive Codes, 187
Introduction, 187
North Carolina Code, 188
South Florida Code, 189
References, 190

Appendix: Basic Concepts on One-Dimensional Vibration 193

Undamped Free Vibration, 193
Damped Free Vibration, 194
Damped Vibration with Sinusoidal
  Forcing Function, 195
Damped Vibration with Random
  Forcing Function, 196
Vibration of Cylinders, 198

Index 199

Foreword

Wind engineering is an emerging discipline. Even though there is a vast amount of literature available in the field of wind engineering, it is diffused in journals and technical reports that relate to the disciplines of structural engineering, fluid dynamics, mechanical engineering, atmospheric sciences, and others. Because of this diffusion, architects and engineers have difficulty in learning and understanding problems associated with wind loads. A simple and easily understandable book can assist practicing professionals in applying the appropriate wind load standard and code provisions in the design of buildings and structures.

Dr. Henry Liu has been involved in the field of wind engineering for the last two decades. His input to the field of wind loads has been significant through research, publications, and committee work. He has provided an outstanding service to the architectural and engineering professions by holding and directing a series of annual short courses on wind effects. Dr. Liu's experience in developing short course notes and his involvement in discussions by participants in the short courses have enabled him to produce this book. His interests and diligence in helping the professional practitioners through this book are commendable.

—*K.C. Mehta*

Preface

Wind is a widespread and costly natural hazard to mankind. Adequate treatment of wind effects in design is essential to the safety and economics of structures. Unfortunately, contemporary college textbooks in structural analysis provide little information on wind effects on structures. The subject is not covered in any other undergraduate course in traditional civil engineering curricula, either. As a result, most practicing structural engineers have inadequate knowledge about wind effects on structures. The few who have adequate knowledge have gained it through no small efforts, such as: spending hundreds of hours struggling through lengthy and difficult-to-read reference books about wind engineering, reading disjointed information contained in various journals and technical publications, and attending pertinent technical conferences and continuing education short courses.

The purpose of this book is to provide important wind engineering knowledge every structural engineer should have for proper understanding of present and future building codes/standards treatment of wind loads, and for proper practice of modern structural engineering. Once the reader has learned the information in this book, he or she will easily understand the treatments of wind loads in all American and international codes and standards, and will have the basic knowledge required for proper comprehension of most technical literature in wind effects on structures.

The contents of this book have been derived from eleven continuing education short courses on wind effects organized and taught, in part by me, at the University of Missouri-Columbia (UMC) during the past 13 years, from a course on wind engineering taught by me at UMC, and from many other sources—reference books, manuals, reports, journals and conference proceedings. I wanted to make this book as useful as possible to the readers. Consequently, an effort has been made to extract the most pertinent information from these sources, and to present such information in an easy-to-understand and easy-to-use format. Many examples are included to illustrate theories, and to show how theories can be used. While most of the symbols in this book conform to those commonly used in wind engineering, some are changed for the sake of simplicity. For example, using the symbols F_1, through F_{10} in Table 5.1 for calculating the along-wind response of buildings instead of using a different Greek letter for each of these ten quantities greatly facilitates the finding of these quantities in the table, and therefore simplifies calculation, especially if it is done on a calculator instead of a computer.

Although this book is aimed at the "average structural engineer," even an expert in wind engineering may find parts of the book informative and useful since there are certain subjects such as wind climate, internal pressure, and the wind load provisions of building codes and standards that are treated in greater detail than in other wind engineering books. The book is recommended for reading by all practicing structural engineers, for use as a supplement to college textbooks in structural analysis, and as the text for short courses in wind effects on structures.

Dr. Kishor C. Mehta, professor of civil engineering, Texas Tech University, reviewed the entire original manuscript and provided valuable comments, suggestions and corrections. Other experts reviewed portions of the original manuscript and provided additional input. These reviewers inicuded: Dr. Grant L. Darkow, professor of atmospheric science, University of Missouri-Columbia, Dr. Ted Stathopoulos, professor and associate director, Center for Building Studies, Concordia University, Canada, Dr. John D. Holmes, principal research scientist, Commonwealth Science and Industrial Research Organization, Australia, Dr. Timothy A. Reinhold, director, Wind Dynamics Laboratory, Raleigh, North Carolina, and Dr. Giovanni Solari, associate professor of civil engineering, University of Genova, Italy.

The manuscript was typed by my sons Jerry and Jason. Jerry, an engineering student, prepared all the preliminary drawings. Finally, it would not have been possible to complete the manuscript smoothly and on schedule without the help and understanding of my wife Dou-Mei (Susie) who, despite her professional career, freed me from most household chores during the months of intense writing. My youngest son Jeffrey also gracefully accepted a long delay in building him a treehouse, pending the completion of the manuscript. This book is dedicated to my mother, Remei Bardina, and my father, Yen-Huai Liu.

1

High Winds and Severe Storms

1.1 INTRODUCTION

Most high winds are produced by severe storms such as hurricanes, tornadoes, thunderstorms, downbursts, and so on. These storms contain not only high winds but also heavy rain and sometimes even hail. High moisture coupled with a warm surface air is conducive to severe storms. Some high winds such as the mountain downslope wind are caused by topographical effects rather than by severe storms. A brief discussion of various types of high winds is provided next.

1.2 TYPES OF HIGH WINDS

Hurricanes, Typhoons, and Cyclones

Different names. Hurricanes, typhoons, cyclones, and so on are different names for the same type of severe storm occurring in different geographical regions. Those occurring in the United States, including Hawaii (more specifically, those occurring in the North Atlantic, the Caribbean Sea, the Gulf of Mexico, and the Western part of the South

Pacific), are called **hurricanes**. Those encountered in the Far East such as China, Japan, Taiwan, and the Philippines (more specifically, those in the South Sea and the Pacific Northwest) are called **typhoons**. Those affecting India and Australia (more specifically, those in the Indian Ocean, Arabian Sea, and Bay of Bengal) are called **cyclones**. They are also under various other names in non–English-speaking countries. In the ensuing discussion, they will all be referred to as **hurricanes**.

Origin, movement, and life span. Hurricanes are generated by low-pressure centers above the ocean at low (5- to 20-degree) latitudes. They move away from the equatorial regions toward higher latitudes. The energy that feeds hurricanes is the latent heat released from condensation of the moisture contained in hurricanes. As the moisture in a hurricane is lost through rain, new moisture is fed into the hurricane due to intense evaporation from the ocean caused by the low pressure and high wind in the hurricane. This mechanism perpetuates and even strengthens hurricanes over the ocean. However, as soon as a hurricane has reached land, it dies down due to lack of moisture and increased surface resistance to wind. Therefore, hurricane winds are strong only over the ocean and in adjacent coastal areas (within approximately one hundred killometers of coastlines). The life span of a hurricane is of the order of one to three weeks, averaging about 10 days.

Direction of rotation. A hurricane is a large body of rotating air. Due to the Coriolis force generated by the Earth's rotation, hurricanes in the Northern Hemisphere (e.g., hurricanes encountered in the United States and typhoons in the Far East) always rotate in the counterclockwise direction. In contrast, hurricanes in the Southern Hemisphere (such as the cyclones in Australia) rotate clockwise.

Translational speed. The **translational speed**, also called the **storm speed**, is the speed at which the center of a hurricane moves. This should not be confused with the **wind speed** in the hurricane. The latter is often much higher than the former. The translational speed of hurricanes can be anywhere between 0 and 100 km/hr. Normally, it is between 10 and 50 km/hr.

Shape, size, and structure. A hurricane is a large funnel-shaped storm with a wide top of the order of 1000 km in diameter and a narrow bottom of the order of 300 to 500 km in diameter. The height of the storm is of the order of 10 to 15 km.

The diameter of a hurricane, encompassing the region of relatively strong wind, is of the order of 500 km. The center part of a hurricane having a diameter of the order of 30 km is called the **eye**; the boundary of the eye is called the **wall**. The eye is a region of clear to partly cloudy skies absent of rain and strong winds. The wall is a region packed with high winds and intense rain. While rain falls in the inner region of the wall, warm humid air rises in the outer part of the wall to supply energy to the hurricane.

Flow pattern. The air outside a hurricane eye circles around the eye and spirals inward at low heights with increasing speed toward the eye. Upon reaching the wall, the air rushes upward to large heights. Then, it spirals outward from the upper region of the hurricane.

Wind speed and pressure distribution. The horizontal distribution of wind speed in a hurricane can be approximated by the Rankine vortex model as follows

$$V_\theta = \frac{V_R r}{R} \qquad \text{(for } r < R\text{)} \tag{1.1}$$

$$V_\theta = \frac{V_R R}{r} \qquad \text{(for } r > R\text{)} \tag{1.2}$$

where V_θ is the tangential (circumferential) component of the wind velocity in a hurricane relative to the movement of the center of the hurricane, r is the radial distance from the hurricane center, and R is the radial distance to the place of maximum velocity, V_R. Note that wind speeds V_θ and V_R refer to the upper-level (gradient height) wind.

The velocity distribution in a hurricane as given by Eqs. 1.1 and 1.2 is depicted in Figure 1.1(a). From Figure 1.1(a), the wind speed in a hurricane reaches a maximum at a distance R from the center, where R corresponds to the radius of the eye. The speed decreases rapidly and linearly from V_R at R to zero toward the center; it decreases more gradually outward in the region $r > R$. The region of the fastest change of wind speed is in the wall ($r \approx R$).

The Rankine vortex model yields the following pressure distribution in a hurricane:

$$p = p_c + \frac{\rho r^2 V_R^2}{2R^2} \qquad \text{(for } r < R\text{)} \tag{1.3}$$

$$p = p_R + \frac{\rho V_R^2}{2}\left(1 - \frac{R^2}{r^2}\right) \qquad \text{(for } r > R\text{)} \tag{1.4}$$

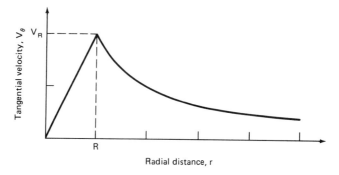

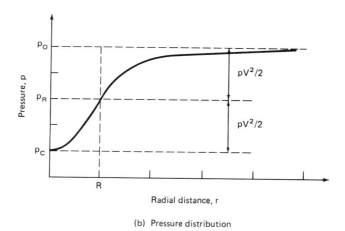

Figure 1.1 Horizontal distributions of wind speed and pressure in a hurricane or tornado according to Rankine Vortex model.

where p is the pressure at radius r of the hurricane, p_c is the pressure at the center of the hurricane, p_R is the pressure at R where the hurricane wind speed is a maximum, V_R, and ρ is the density of air.

From Eqs. 1.3 and 1.4, the pressure p_R at R and the pressure p_o outside the hurricane are, respectively,

$$p_R = p_c + \frac{\rho V_R^2}{2} \quad \text{(for } r = R\text{)} \tag{1.5}$$

$$p_o = p_R + \frac{\rho V_R^2}{2} \quad \text{(for } r >> R\text{)} \tag{1.6}$$

Sec. 1.2 Types of High Winds 5

The foregoing pressure distribution in a hurricane is illustrated in Figure 1.1(b).

From the foregoing, it can be seen that the pressure in a hurricane is a minimum at the center, rising with radial distance r. The pressure p_R at R is higher than the pressure p_c at the center by an amount equal to the dynamic pressure (velocity pressure) $\rho V_R^2/2$. Likewise, the pressure p_o outside the hurricane sphere of influence is higher than the pressure p_R by the amount $\rho V_R^2/2$. Therefore, the total increase in pressure from the center to the outside is ρV_R^2 or twice the dynamic pressure. This pressure distribution is shown in Figure 1.1(b).

Example 1-1. A coastal weather station measured a temperature of 85°F and an atmospheric pressure of 1000 mb just before the station was affected by a hurricane. What would the pressure at the station be when the center of the hurricane reaches the station? The maximum wind speed in the hurricane measured by an aircraft flying at 3000 m height was 200 mph.

[Solution]. Using $T = 85°F = 460 + 85 = 545°R$ and $p = 1000$ mb $= 10^5$ Pa $= 2090$ psf, the equation of state of perfect gas gives the density of the air as

$$\rho = \frac{p_o}{ET} = \frac{2090}{1715 \times 545} = 0.00224 \text{ slugs/ft}^3$$

As the center of the hurricane passes the station, the pressure drop is expected to be

$$p_o - p_c = \rho V_R^2 = 0.00224 \times (200 \times 1.467)^2 = 192.8 \text{ psf} = 92 \text{ mb}$$

Therefore, the pressure at the station when the center of the hurricane reached the station is expected to be

$$p_c = p_o - 92 \text{ mb} = 1000 - 92 = 908 \text{ mb}$$

Surface wind. As in the case of most other types of winds, the wind speed in a hurricane decreases with decreasing height, reaching zero velocity at ground level to satisfy the no-slip condition of fluid mechanics. However, what is normally referred to as the **surface wind** is the wind speed not **at** the surface but rather **near** the surface, measured by anemometers mounted normally at a height of 10 m (33ft) above ground. When using hurricane wind data, one should distinguish between surface wind speed and the wind speed measured by aircraft at

TABLE 1.1 Classification of Wind by Beaufort Scale

Beaufort Number	Wind Speed (mph)	Descriptor	Effect observed
0	0–1	Calm	Smoke rises vertically
1	2–3	Light air	Smoke drifts; vanes do not move
2	4–7	Light breeze	Leaves rustle; vanes begin to move.
3	8–12	Gentle breeze	Leaves in constant motion; light flags extended.
4	13–18	Moderate breeze	Dust, leaves raised; small branches move.
5	19–24	Fresh breeze	Small trees begin to sway
6	25–31	Strong breeze	Large branches of trees in motion; whistling heard in wires.
7	32–38	Near gale	Whole tree in motion; resistance felt in walking against wind.
8	39–46	Gale	Twigs and small branches break; progress generally impeded.
9	47–54	Strong gale	Slight structural damage occurs; slate blown from roofs.
10	55–63	Storm	Trees broken or uprooted; considerable structural damage occurs.
11	64–73	Violent storm	Damage all over
12	74 and above	Hurricane	Large-scale damage, calamity

high altitudes, for the latter may be considerably higher than the former. Except for very tall structures, the surface wind speed is what is encountered by structures.

Intensity ratings. Meteorologists do not call a hurricane-type storm a hurricane unless or until the maximum wind speed of the storm has reached 75 mph (33.5 m/s). At lower speeds, the storm is called a **tropical storm**. The famous Beaufort scale, suggested by English Admiral Beaufort in 1806 and slightly modified since then, gives hurricanes the highest rating number, 12, as seen in Table 1.1. In Japan, a typhoon

TABLE 1.2 Classification of Hurricanes by Saffir/Simpson Scale

Saffir/Simpson Number	Wind Speed (mph)	Possible Tidal Surge (feet above sea level)	Damage Potential
1	74–95	4–5	Minimal
2	96–110	6–8	Moderate
3	111–130	9–12	Extensive
4	131–155	13–18	Extreme
5	156 and above	19 and above	Catastrophic

Note:
1. The wind speed referred to in the Saffir-Simpson scale is the "sustained speed" at 10-m height. The U.S. National Weather Service uses 1-minute average to determine the "sustained wind speed."
2. The damage to houses are for those in hurricane regions which normally follow more stringent construction practices than in the nonhurricane regions.

Scale no. 1. Damage primarily to shrubbery, trees, foliage, unanchored mobile homes, and so on.

Scale no. 2. Major damage to exposed mobile homes. Extensive damage to poorly constructed signs. Some damage to roofing materials, windows, and doors.

Scale no. 3. Large trees blown down, some structural damage to small homes, and so on.

Scale no. 4. All signs blown down, extensive damage to roofing materials, windows and doors, complete failure of roofs on many houses, complete destruction of mobile homes, and so on.

Scale no. 5. Complete failure of roofs on many industrial buildings, small buildings overturned or blown away, and so on.

is not called a typhoon unless its wind speed has reached a Beaufort scale of 8. The U.S. National Weather Service uses the **Saffir/Simpson scale** to rate the intensity and the damage potential of hurricanes—see Table 1.2.

Causes of damage. Hurricane damage is caused not just by high winds. Surges of the sea (storm surges) and the action of strong waves also pose severe threat to lives and properties in low areas along coastlines. For instance, the November 12, 1970 cyclone that hit Bang-

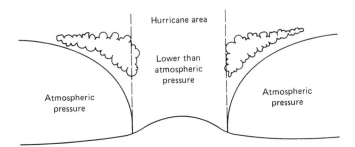

Figure 1.2 Surges caused by the low pressure of hurricane (equilibrium under hydrostatic condition causes sea to surge where pressure is lower than atmospheric).

ladesh (what used to be East Pakistan) killed almost half a million people—mostly washed into the sea by surges and water waves. It was the worst hurricane disaster in history. Hurricanes also bring torrential rain which often causes severe flooding problems. Therefore, design against hurricanes must consider not only high winds but also the hurricane-generated surges, waves, and floods.

Storm surges less than 1 m in height can be produced by the suction of the low-pressure centers of hurricanes; see Figure 1.2. Larger surges can be produced by the strong wind of hurricanes when the wind direction is onshore. This is called **wind tide** or **setup**. When a wind tide coincides with a normal tide[1] at a given location, large surges can be produced. This is especially serious in certain gulfs and estuaries where the topography is conducive to tidal surges. Finally, large water waves are generated by hurricane winds. Figure 1.3 illustrates the combined effect of wind tide and waves.

Example 1-2. Estimate the maximum surge of ocean water that can be generated by the low pressure center of the hurricane described in the previous example.

[Solution]. From equation of hydrostatics, the maximum surge of ocean that can be generated by the low pressure center of the storm is $H = \Delta p/\gamma$, where Δp is the pressure difference between the center of the hurricane and a point outside the storm, and γ is the specific weight of the sea water, assumed to be 64 lb/ft.^{3} Since $\Delta p = p_o - p_c = 193$ psf, $H = 193/64 = 3.02$ ft $= 0.919$ m.

---

[1] Ordinary tides are produced by the change of gravitational forces exerted by the moon and the sun as the earth rotates.

Sec. 1.2   Types of High Winds   9

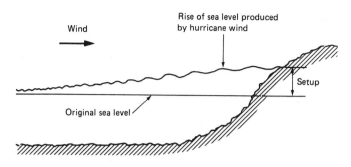

Figure 1.3   Waves and tides generated by hurricane wind.

This example shows that even in such a strong hurricane, the storm surge caused by hydrostatic effect is only 3 ft. It shows that the large storm surges normally accompanying hurricanes must have been caused primarily by wind tide or setup rather than by the low pressure that exists in hurricanes.

Season and frequency.   Hurricanes occur most frequently in late summer when the ocean water temperature has reached a maximum. In the Northern Hemisphere, this means August–September, and in the Southern Hemisphere, this means February–March. The winter is almost entirely absent of hurricanes; see Figure 1.4.

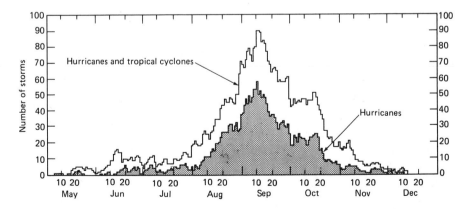

Figure 1.4   Monthly variation of hurricanes and tropical cyclones of the north Atlantic Ocean. (From Neumann et al. 1981).

Maximum speed. The maximum wind speeds of hurricanes and tornadoes have both been grossly overestimated in the past. Based on severe damages caused by such storms, many meteorolgists and engineers used to think that tornadoes could have wind speed approaching or even exceeding the speed of sound, and that hurricanes could have surface wind speed higher than 100 m/s (224 mph). Such myths are still being perpetuated in some recent publications (e.g., Nalivkin 1982). However, it is now widely held by experts that tornado wind speed may never exceed 300 mph (137 m/s), and hurricane surface wind speed may never exceed 200 mph (90 m/s). Some of the maximum hurricane surface wind speeds ever measured are 85 m/s (189 mph) peak gust at the Muroto Weather Station in Japan, and 75 m/s (167 mph) fastest 10-min wind at Lan-Yu (Orchard Island) in Taiwan. Both stations are located on high hilltops which amplify winds.

Tornadoes

Cause. Both severe thunderstorms and hurricanes can generate tornadoes, with the former being the more frequent cause. It is generally accepted by meteorologists that severe thunderstorms are often associated with a surface-based warm and moist air layer approximately 1 km in depth, a dry cool layer above centered at about 2–4 km, and above this a high-speed **jet stream** centered at 10–12 km. The interaction of the jet stream with the dry layer cool air and the ascending warm moist air often produces a sinking mass of evaporatively cooled air which, when conditions are ripe, generates tornadoes and downbursts.

Geographical distributions. Tornadoes are found in all parts of the world, with the United States being the country most frequented and plagued by tornadoes. The Midwest (i.e., the central region of the United States), especially Oklahoma and its neighboring states, has the greatest number of tornadoes. This region is sometimes referred to as **the tornado belt.** Although tornadoes have occurred in certain localities (cities and counties) far more frequently than in others, the variation of tornado frequency within small geographical areas is statistically insignificant. For instance, the fact that Codell, Kansas, was hit by a tornado three years in a row (1916, 1917, 1918) and on the same day of each year (May 20) is a mere coincidence.

Tornadoes can hit any area or region of the United States. For instance, the city of Los Angeles, California, was hit by a damaging tornado on March 1, 1983. Prior to that date, few individuals thought that a tornado would ever occur in Los Angeles.

Shape, size, and detection. A tornado has the same funnel shape as a hurricane except that a tornado is much more slender (smaller in horizontal dimension). The largest tornadoes (they also happen to be the strongest and most devastating) have a diameter of the order of 1 km. Most tornadoes have a diameter (judged by damage areas) smaller than 400 m. Due to the slenderness and comparatively short life span of tornadoes, they are much more difficult to detect than are hurricanes. Consequently, many tornadoes did not show up in the statistics of tornadoes. However, since 1953, improvements in tornado observation and reporting techniques have drastically reduced the number of unreported tornadoes. This trend is revealed in tornado statistics showing a steady increase in tornado frequency in recent years, especially since 1953 (Darkow 1986). The increase may also be caused in part by the tendency of news media in attributing to tornadoes for wind damages caused by unknown types of wind storms.

Direction of rotation. Unlike hurricanes whose direction of rotation (counterclockwise in the Northern Hemisphere and clockwise in the Southern Hemisphere) is dictated by the Coriolis force, this force has a much weaker effect on tornadoes—due to the small size of tornadoes. Consequently, most but not all tornadoes in the Northern Hemisphere rotate counterclockwise.

Path, translational speed, and life span. The paths of most tornadoes are straight or gently curved. Since most tornadoes are associated with thunderstorms, they move normally in the same direction and at about the same translational speed as their parent storms. In the Midwest, tornadoes move generally from the southwest toward the northeast, at an average translational speed of the order of 50–60 km/hr. The instantaneous translational speed may be anywhere between 0 and 100 km/hr. In contrast to hurricanes, which have a life span in the neighborhood of 10 days, and a path length greater than a thousand miles, most tornadoes last no more than 30 minutes (many disappear within minutes), with a path length of less than 25 km. Only the strongest tornadoes—those happen rarely—have a path of over 100 km and a life span longer than an hour. For example, the infamous Tri-State Tornado that hit Missouri, Illinois, and Indiana on March 18, 1925, and that killed close to 700 people had a continuous path of 353 km, an average width of 1200 m, and a life span of 3.5 hours. The translational speed of this tornado was 100 km/hr.

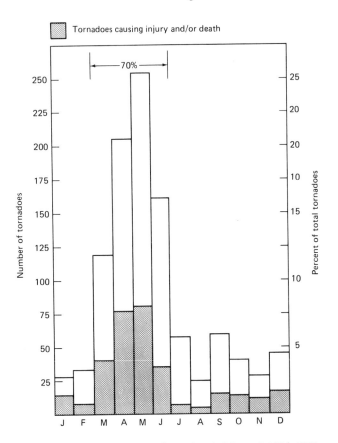

Figure 1.5 Monthly variation of tornadoes in Missouri, 1916–1980. (From Darkow, 1986).

Seasonal variation. Although tornadoes can occur anytime during a year, they are most frequent during certain months—the so-called **tornado season**. The tornado season differs for different geographical regions. In the United States, the tornado season has a general tendency of occurring later as one goes northward. For instance, in Alabama, Louisiana, and Mississippi, tornadoes happen most frequently in March and April. In Missouri and Kansas, the peak month of tornadoes is shifted to May, and in North Dakota and Minnesota, it is June. This general trend of tornado activities drifting northward as time of year progresses is due to the northward movement of a low-level tongue of warm moist air originating from the Gulf of Mexico, and a concurrent northward drift of jet streams. The seasonal variation of tornadoes in Missouri is shown in Figure 1.5.

Sec. 1.2   Types of High Winds

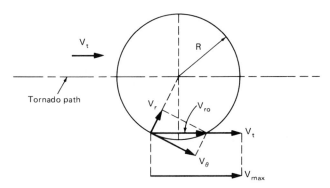

Figure 1.6   Combination of horizontal components of velocity in a tornado.

Velocity fields. The velocity in a tornado relative to the motion of the center or axis of the tornado can be expressed in cylindrical coordinates as follows:

$$\vec{V}_{rel} = V_r \hat{r} + V_\theta \hat{\theta} + V_z \hat{z} \qquad (1.7)$$

where $\vec{V}_{rel}$ is the relative velocity; $\hat{r}$, $\hat{\theta}$, and $\hat{z}$ are unit vectors in the radial, tangential, and axial directions, respectively; and V_r, V_θ, and V_z are three velocity components in the r, θ, and z directions, respectively.

The tangential component V_θ of a tornado can be modeled by the same Rankine vortex for hurricanes discussed before. The vector sum of the two horizontal orthogonal components V_r and V_θ is often referred to as **rotational component** V_{ro}, namely,

$$V_{ro} = \sqrt{V_r^2 + V_\theta^2} \qquad (1.8)$$

The horizontal velocity of tornado winds encountered by a stationary object or person is the combination of the rotational component V_{ro} and the translational speed of the tornado, V_t. As shown in Figure 1.6, the maximum horizontal velocity V_{max} occurs at a place where V_{ro} and V_t are both in the same direction, namely,

$$V_{max} = V_{ro} + V_t \qquad (1.9)$$

Pressure field. The pressure field in a tornado is the same as that given by Eqs. 1.3 through 1.6 for hurricanes. In fact, these equations are more accurate for tornadoes than for hurricanes, because the smaller diameter of tornadoes results in a diminished effect of the Coriolis force on tornadoes.

Atmospheric pressure change. When a tornado approaches a stucture, a rapid drop of atmospheric pressure with time is encountered by the structure. If the structure is enclosed as in the case of a building with its door and windows shut, the internal pressure of the structure will remain high as the external pressure decreases. This develops a large difference between internal and external pressures, causing the structure to explode. This phenomenon can be analyzed as follows:

Consider a tornado moving along a straight-line path with translational speed V_t as shown in Figure 1.7. Suppose the tornado is closest to the building when $t = 0$, and it leaves the building when $t > 0$. From geometry,

$$r^2 = x^2 + y^2 = V_t^2 t^2 + y^2 \tag{1.10}$$

Substituting Eq. 1.10 into Eqs. 1.3 and 1.4 yields

$$p = p_c + \frac{\rho(V_t^2 t^2 + y^2) V_R^2}{2R^2} \qquad \text{(for } |t| < \frac{\sqrt{R^2 - y^2}}{V_t}\text{)} \tag{1.11}$$

$$p = p_R + \frac{\rho V_R^2}{2}\left(1 - \frac{R^2}{V_t^2 t^2 + y^2}\right) \qquad \text{(for } |t| > \frac{\sqrt{R^2 - y^2}}{V_t}\text{)} \tag{1.12}$$

If the building is on the path of the tornado, then $y = 0$ and the last two equations become

$$p = p_c + \frac{\rho V_R^2 V_t^2 t^2}{2R^2} \qquad \text{(for } |t| < \frac{R}{V_t}\text{)} \tag{1.13}$$

$$p = p_R + \frac{\rho V_R^2}{2}\left(1 - \frac{R^2}{V_t^2 t^2}\right) \qquad \text{(for } |t| > \frac{R}{V_t}\text{)} \tag{1.14}$$

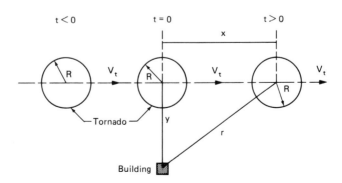

Figure 1.7 Analysis of atmospheric pressure change caused by an approaching tornado.

Sec. 1.2 Types of High Winds

The foregoing equations can be used for both approaching tornadoes ($t < 0$) and departing tornadoes ($t > 0$). They give the external pressure encountered by the building as a function of time.

The variation of the internal pressure with time can be obtained by treating the leakage flow through the building as flow through an orifice. By using the equation of flow through an orifice, the continuity equation based on mass balance, and the equation of state for perfect gas, it can be proved that during the period of decompression (i.e., when $p_i > p$),

$$\frac{dp_i}{dt} + \left(\frac{C_d A}{\mathcal{V}} \sqrt{2ET_i}\right) \sqrt{(P_i - p)p_i} = 0 \qquad (1.15a)$$

where p_i is the internal pressure; c_d is the orifice discharge coefficient which is of the order of 0.7; A is the total area of the opening on the building; $\mathcal{V}$ is the internal volume of the building; E is the engineering gas constant; T_i is the temperature of the air inside the building, assumed constant with time; and p is the external pressure given by Eqs. 1.13 and 1.14. Note that all the pressures used in the equations, p_i and p, must be absolute pressure, and T_i must be absolute temperature.

On the other hand, during the subsequent period of compression (i.e., when $p > p_i$), we have

$$\frac{dp_i}{dt} - \left(\frac{C_d A}{\mathcal{V}} \sqrt{2ET_i}\right) \sqrt{(p - p_i)p_i} = 0 \qquad (1.15b)$$

Equations 1.15a and 1.15b can be solved numerically by using the initial condition $p_i = p$ when $t = \pm\infty$. In practice, computation can start at a finite time such as $t = -5R/V_t$, with an assumed value of p_i slightly higher than p, such as $p_i = p_o = p_c + \rho V_R^2$.

Example 1-3. A tornado has a maximum wind speed $V_R = 200$ mph at $R = 500$ ft. The tornado is heading toward a house at a translational speed of $V_t = 50$ mph. The house has a total internal volume of 50,000 ft³ and a total opening area of 1.0 ft². The air temperature is 85°F, and the atmospheric pressure outside the tornado is $p_o = 1000$ mb. Find the maximum load on the house due to the atmospheric pressure change.

[Solution]. Since the temperature, pressure, and maximum wind speed for this case are the same as in Example 1-1, from the previous example $\rho = 0.00224$ slugs/ft³, $p_c = 908$ mb $= 1,898$ psf, $V_R = 200$ mph $= 293$ fps, $\rho V_R^2 = 192.8$ psf, and $p_R = p_c + 0.5\rho V_R^2 =$

1994 psf. Furthermore, $V_t = 50$ mph $= 73.4$ fps and $R = 500$ ft. Therefore, Eqs. 1.13 and 1.14 reduce to, respectively,

$$p = 1898 + 2.08t^2 \qquad (|t| < 6.81) \qquad \text{(a)}$$

$$p = 2090 - \frac{4473}{t^2} \qquad (|t| > 6.81) \qquad \text{(b)}$$

Equations (a) and (b) yield the solid line in Figure 1.8.

With $C_d = 0.7$, $A = 1.0$ ft^2, $V = 50,000$ ft^3, and $E = 1715$ ft-lb/slug°R, Eqs. 1.15 a and b become, respectively,

$$\frac{dp_i}{dt} + 0.01914 \sqrt{(p_i - p)p_i} = 0 \qquad \text{(c)}$$

and

$$\frac{dp_i}{dt} - 0.01914 \sqrt{(p - p_i)p_i} = 0 \qquad \text{(d)}$$

Solution of Eqs. (c) and (d) over the time period from $t = -40$ s to $+40$ s yields the lowest dashed line in Figure 1.8 for internal pressure.

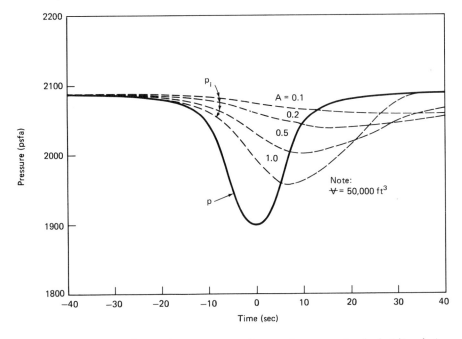

Figure 1.8 Variation of internal pressure, p_i, and external pressure, p, of a building during tornado passage. (Note that time is negative before tornado reaches the building and positive thereafter.)

Sec. 1.2   Types of High Winds   17

The maximum bursting pressure on the building for this case is $(\Delta p)_{max} = (p_i - p)_{max} = 104$ psf. For this example, $(\Delta p)_{max}$ occurs at approximately 2 seconds before the center of the tornado reaches the building.

In a similar manner, the variations of internal pressure with time for this building at three smaller openings were calculated and plotted in Figure 1.8. It is clear from Figure 1.8 that as the opening gets larger, the internal pressure p_i approaches the external pressure p.

Fujita scale. In 1971, T. Fujita at the University of Chicago proposed a scale to determine the maximum wind speed in any tornado from the damage observed. The scale, now routinely used by the National Weather Service, is generally referred to as the **Fujita scale** or **F-scale**. The scale is divided into 13 numbers: F-0 through F-12. Larger F-scale numbers represent higher speeds, with F-12 reaching sonic velocity (Mach 1). The various F-scale numbers and the corresponding wind speed and damage are shown in Table 1.3.

The F-scale is a simple method for determining the wind speed of each tornado based on the damage observed. It serves the National Weather Service well for it allows a quick estimate of the maximum wind speed of each of the hundreds of tornadoes occurring every year in the United States. However, caution must be exercised in using or interpreting the F-scale for it contains the following problems:

1. While the wind speeds given for the lower scales such as F-0 and F-1 appear to be approximately correct, the speeds given for the higher scales (e.g., F-4 and F-5) appear exaggerated.
2. Recent research findings indicate that it is quite questionable that any tornado can have a wind speed exceeding 318 mph. Therefore, F-6 through F-12 should be dropped from the scale to avoid confusion or misuse.
3. Recent studies indicate that the F-scale has not been used consistently. What one state classifies as an F-0 tornado may be classified as F-1 and occasionally even F-2 in another state. This is due to the vague and incomplete description of the damage associated with each F-number.
4. The F-scale does not reflect the variation of construction practice. For instance, the construction of houses in the coastal areas of Florida and North Carolina is governed by special building codes that require hurricane-resistant construction. The houses constructed there can resist a much higher wind than the houses

constructed in most of the other parts of the nation. Therefore, what is classified as an F-0 tornado in a hurricane-prone region may produce an F-2 damage in another region having a weak building code or no codes at all.

Due to the foregoing problems, it is expected that the Fujita scale will be revised by the National Weather Service in the future.

TABLE 1.3 Classification of Tornadoes by Fujita Scale

Wind Speed	Damage Potential	Effect Observed
F-0 (40–72 mph)	Light	Some damage to chimneys; branches broken off trees; shallow-rooted trees pushed over; sign boards damaged.
F-1 (73–112 mph)	Moderate	Peel surface off roofs; mobile homes pushed off foundations or overturned; moving autos pushed off roads.
F-2 (113–157 mph)	Considerable	Roof torn off from houses; mobile homes demolished; boxcars pushed over; large trees snapped or uprooted, light-object missiles.
F-3 (158–206 mph)	Severe	Roofs and some walls torn off well-constructed houses; trains overturned; most trees in forests uprooted; heavy cans lifted off ground and thrown.
F-4 (207–260 mph)	Devastating	Well constructed houses leveled; structures with weak foundations blown off some distance; cars thrown and large missiles generated.
F-5 (261–318 mph)	Incredible	Strong frame houses lifted off foundations and carried considerable distance to disintegrate; automobile-sized missiles fly through air in excess of 100 m; trees debarked; incredible phenomena will occur.
F-6 to F-12 (319 mph–sonic velocity)	Currently believed nonexistent.	

Tornado like vortices. A **waterspout** is identical to a tornado except that it is on water and hence sucks water into the air. As soon as a waterspout has reached land and maintains itself, it is called a **tornado**. A **dust devil** is a tornadolike vortex occurring over dry surfaces during periods of intense solar heating such as in desert areas. Unlike a tornado, a dust devil contains no moisture and is not associated with any parent cloud or thunderstorm. Its wind speed is normally not more than 40 mph.

A large tornado sometimes contains several satellite vortices circulating around the tornado center; they are called **suction vortices.** Due to their high translational speed (the speed that they circulate around the tornado center) and an additional rotational speed (the speed they spin around their own axes), suction vortices can generate wind speeds higher than that of their parent tornado. They are evidenced by cycloidal (helical) ground marks observed after the passage of certain tornadoes.

Tornado design criteria. The U.S. Nuclear Regulatory Commission (NRC) requires that all safety-related structures in a nuclear power plant be designed against the strongest possible tornadoes. The commission has divided the continental United States into three tornado regions as shown in Figure 1.9. The design wind speeds and the atmo-

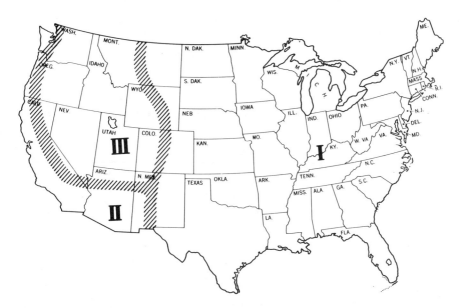

Figure 1.9 Classification of tornado intensity regions by U. S. Nuclear Regulatory Commission. (From Markee, Beckerley, and Sanders, 1974.)

spheric pressure changes for each region are specified in Table 1.4. Note that in Region I, which is the most tornado-intensive region, the required design maximum wind speed is 360 mph (161 m/s), the required translational speed of the tornado is 70 mph (31.3 m/s), the required pressure drop ($p_o - p_c$) is 3.0 psi (432 psf or 20.7 kPa), and the required rate of pressure drop is 2 psi/s (288 psf/s or 13.8 kPa/s). More about NRC tornado design criteria for nuclear power plants is given in Abbey (1975). A somewhat different standard has been promulgated by the American National Standard Institute (ANSI/ANS 1983). The maximum wind speed in this standard is 320 mph (143 m/s), and the maximum required pressure drop is 1.96 psi (282 psf or 13.5 kPa).

Thunderstorm Winds, Straight-Line Winds, and Downbursts

Thunderstorm and straight-line winds. Severe thunderstorms generate high winds and sometimes even tornadoes. The nonspinning (nontornadic) types are often referred to as **thunderstorm wind** or **straight-line wind**. Although straight-line winds are normally not as intense as tornadoes, they produce far more accumulative damage than tornadoes because they occur far more frequently. This is true even in tornado-prone areas such as Illinois or Oklahoma. Straight-line winds can have speeds approaching or sometimes exceeding 100 mph (44.7 m/s), causing roofs to be blown off; mobile homes, automobiles, and parked aircraft overturned; trees toppled; power lines downed; and so on.

Downbursts. A particular type of thunderstorm wind, called a **downburst**, is generated by a falling mass of evaporatively cooled air frequently driven by hail and heavy rain in the parent thunderstorm. As

TABLE 1.4 U.S. Nuclear Regulatory Commission Required Design Basis Tornado Characteristics

Tornado Region	V_{max} (mph)	V_{ro} (mph)	V_t (mph) max.	V_t (mph) min.	R (ft)	$p_o - p_c$ (psi)	$\dfrac{dp}{dt}$ (psi/s)
I	360	290	70	5	150	3.0	2.0
II	300	240	60	5	150	2.25	1.2
III	240	190	50	5	150	1.5	0.6

this falling air mass impinges on ground, it spreads out horizontally and generates strong surface winds of short duration. The situation is analogous to the flow generated by pouring water on ground from a pail mounted on a moving truck, with the parent storm being the moving truck.

T. Fujita (1985) classified downbursts into two size groups: **microbursts** and **macrobursts**. A microburst has a small horizontal scale of the order of a few hundred meters, and has damaging winds lasting from 2 to 5 minutes. On the other hand, a macroburst covers a larger area of the order of 1–5 km, and the damaging winds last 5 to 30 minutes. A downburst may be moving or stationary. The streamlines in a downburst can be straight or curved. A curved downburst may sometimes develop into a tornado.

A classical example of damage caused by a downburst is the roof failure of the Kemper Arena in Kansas City, June 4, 1979. While the collapse of the roof of this large structure cost more than $5 million, luckily no one was under the roof at the time of the collapse. Had there

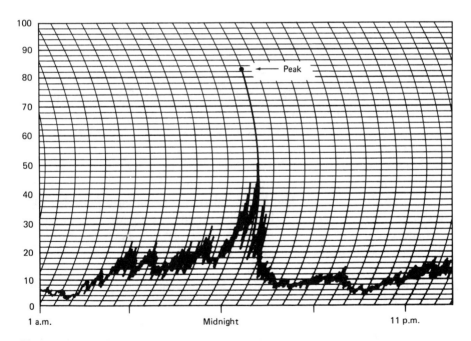

Figure 1.10 Wind speed record of a microburst at Columbia Regional Airport, Missouri, June 17, 1985 The sharp peak at midnight reads 83 knots which is equivalent to 96 mph or 43 m/s. (Courtesy of U.S. National Weather Service.)

Figure 1.11 Damage to parked aircraft by the microburst that hit Columbia Regional Airport, Missouri, June 17, 1985. (Courtesy of U. S. National Weather Service).

been an event scheduled in the arena at the time of the storm, it could have been a major national disaster.

A salient characteristics of microbursts is a sharp peak (spike) in the wind speed chart recorded at any station hit by a microburst. For instance, the chart in Figure 1.10 shows a sudden rise of wind speed from about 40 mph to 96 mph within a few seconds. This wind speed record is for a microburst that hit the Regional Airport, Columbia, Missouri, at midnight on June 17, 1985 (Liu and Nateghi 1988). Among the various damage caused were 24 demolished aircraft; see Figure 1.11. Fujita (1985) gives the criteria for identifying downbursts.

Mountain downslope winds. Mountain downslope winds happen when a cold layer of air descends from the peak of a mountain in a manner similar to water flowing down a steep slope. Due to the acceleration caused by gravity, the wind reaching the foothill can gain speeds as high as those of hurricanes. The phenomenon is referred to in scientific

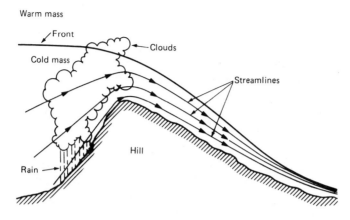

Figure 1.12 Mountain downslope wind.

literature as the **mountain downslope winds**. Note that for any air mass to be able to accelerate by gravity, the air must be a cold layer under a warm upper air, such as that can be generated by a cold front. This type of wind often occurs in winter when a cold front crosses a mountain; see Figure 1.12.

As the cold air in a mountain downslope wind descends down a mountain, not only does the wind speed increase, but the air temperature also rises due to adiabatic compression of the air caused by increasing hydrostatic pressure encountered at lower elevations. For this reason, mountain downslope winds often bring warmer temperature to low areas.

Mountain downslope winds occurring in different geographical regions are under different names. In the Rocky Mountains areas of the United States they are called **chinook**, in southern California they are called the **Santa Ana wind**, in the Alps of Europe they are called **foehn**, and in Yugoslavia they are called **bora**. Note that *foehn* and *bora* are now being used as generic terms for warm and cold mountain downslope winds, respectively. Many communities on the eastern slope of the Rocky Mountains are plagued by mountain downslope winds. For example, Boulder, Colorado, each year experiences more than one downslope wind of a speed exceeding 100 mph, sometimes approaching 130 mph.

REFERENCES

ABBEY, R. F. (1975). "Establishment of Maximum Regional Tornado Criteria for Nuclear Power Plants," *Proceedings of the 9th Conference on Severe Local Storms,* American Meteorological Society, 368–375.

ANSI/ANS-2.3 (1983). *American National Standard for Estimating Tornado and Extreme Wind Characteristics at Nuclear Power Sites,* American Nuclear Society, La Grange Park, Illinois.

DARKOW, G. L. (1986). "Tornado Wind Probabilities for Engineers," Course Notes, 11th Annual Continuing Education Short Course on Wind Effects on Buildings and Structures, University of Missouri-Columbia.

FUJITA, T. T. (1985). *The Downburst,* SMRP Research Paper No. 210, University of Chicago, Chicago, 122 pages.

LIU, H. AND NATEGHI, F. (1988). "Wind Damage to Airport: Lessons Learned," *Journal of Aerospace Engineering,* ASCE, 1(2), 105–116.

MARKEE, E. H., BECKERLEY, J. G. AND SANDERS, K. E. (1974). *Technical basis for Interim Regional Tornado Criteria,* U.S. Nuclear Regulatory Commission, Washington, D.C.

NALIVKIN, D. V. (1982). *Hurricane, Storms and Tornadoes,* P.2 and P.4, translated from Russian by B. B. Bhattacharya, Amerind Publishing Co., New Delhi, India.

NEUMANN C. J. ET AL. (1981). *Tropical Cyclones of the North Atlantic Ocean, 1871–1980,* National Climatic Center, Ashville, North Carolina.

2

Extreme Wind Probability

2.1 INTRODUCTION

Modern design of structures is based on the concept of probability. A structure is designed to provide a specific degree of safety against high winds, determined by the probability of occurrence of high winds with speeds exceeding the design value. For instance, the U.S. national standards on structural loads (ASCE 7-88, formerly ANSI A58.1) requires that ordinary structures be designed for an annual exceedance probability of 2% (equivalent to 50-year return period). The wind speed corresponding to such a probability for any location in the United States is given in Figure 7.1. For crucial facilities such as hospitals, they must be designed according to an exceedance probability of 1% (100-year return period), and for structures of low risk to human life, the design can be based on 4% probability (25-year return period). The purpose of this chapter is to provide the pertinent information needed for understanding this probability concept in wind load determination and to assess the probability of hurricanes, tornadoes, and other high winds.

2.2 EXCEEDANCE PROBABILITY AND RETURN PERIOD

The basic probability of high winds needed for structural design is the **exceedance probability** P_E which is the probability that a given wind speed will be exceeded within a one-year period. The reciprocal of exceedance probability is called the **return period** or **recurrence interval**, namely,

$$T_R \text{ (return period)} = \frac{1}{P_E} \quad (2.1)$$

where P_E and T_R can be determined from high wind data as illustrated in the following example:

Consider the annual fastest wind (i.e., the maximum wind speed of each year) for Columbia, Missouri. In a real study, the entire record of the station, more than 40 years, should be used. However, for purpose of illustration, let us use only a 10-year record, between 1950 and 1959. The annual fastest wind recorded for this station, as for most other stations in the United States, is the **fastest-mile wind,** which will be discussed in Chapter 3. The methodology used here for calculating the exceedance probability is applicable not only to the fastest-mile wind but also to gust speed and high winds averaged over other durations such as 10 minutes.

The data for Columbia, Missouri, are arranged in descending order of wind speed as shown in column 2 of Table 2.1. The values of ex-

TABLE 2.1 Calculation of Exceedance Probability for Annual Fastest-Mile Wind Speed in Columbia, Missouri, (1950–1959)

Wind Speed Ranking r	Annual Fastest-Mile Wind V (mph)	Wind Direction	Year of Occurrence	Exceedance Probability P_E
1	63	NW	1952	0.09
2	61	NW	1958	0.18
3	58	SW	1950	0.27
4	58	NW	1951	0.36
5	57	NW	1953	0.45
6	56	NW	1954	0.55
7	56	NW	1956	0.64
8	56	NW	1957	0.73
9	49	NW	1955	0.82
10	49	NW	1959	0.91

ceedance probability listed in the last column of the table were calculated from

$$P_E = \frac{r}{N+1} \quad (2.2)$$

where r is the **rank number** (column 1) and N is the total number of years—10 in this case. Note that $r/(N+1)$ instead of r/N was used to calculate the exceedance probability in order to avoid $P_E = 1$ (certainty) for $r = N$.

The values of V versus P_E in Table 2.1 can be plotted on a probability paper as shown in Figure 2.1. A straight line on a probability paper respresents **normal (Gaussian) distribution**.

2.3 PROBABILITY DENSITY AND DISTRIBUTION FUNCTIONS

Although the prediction of exceedance probability corresponding to any wind speed can be based on the line of best fit shown in Figure 2.1, extrapolation is difficult because the line is usually not straight—non-

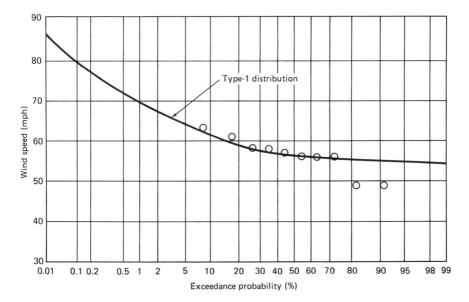

Figure 2.1 Exceedance probability of extreme wind speed (annual fastest-mile wind) for Columbia, Missouri (Note that only 10 years of data, 1950–1959, are used for illustration).

Gaussian. Knowing the correct probability law (probability density and distribution functions) for a given case makes extrapolation more realistic, especially for large values of wind speeds (small probability).

The **probability density function** of wind speed V, designated as $f(V)$, is the probability per unit wind speed. From this definition, the probability for wind speed between any value V and an infinitesimally larger value $V + dV$ is $f(V)\,dV$.

The probability for wind speed to be less than V is

$$F(V) = \int_0^V f(V)\,dV \tag{2.3}$$

from which

$$f(V) = \frac{dF(V)}{dV} \tag{2.4}$$

Note that $F(V)$ is called the **probability distribution function.**

The exceedance probability P_E is related to the probability distribution function by

$$P_E(V) = 1 - F(V) = 1 - \int_0^V f(V)\,dV \tag{2.5}$$

This equation shows that the exceedance probability can be easily calculated from either the probability distribution function or the probability density function.

Different probability laws (models) exist in the literature for the functional forms of $f(V)$ and $F(V)$. While one type of wind data may best be represented by a given probability law, another type of wind data may fit another probability law better. Some most important probability laws for winds and their usages are discussed next.

2.4 PROBABILITY OF ORDINARY WINDS

The wind speed probability for ordinary winds at any location can be approximated by a **Rayleigh distribution** as follows:

$$F(V) = 1 - \exp\left(-\frac{\pi V^2}{4\overline{V}^2}\right) \tag{2.6}$$

where $\overline{V}$ is the mean value of the wind and π is 3.1416.

The Rayleigh distribution is rather simple because it depends only on one parameter—the mean wind speed $\overline{V}$. This distribution is for wind velocity in general (the wind speed that may happen any time at a given location). It should not be used for extreme values such as the fastest annual wind values. Even for wind in general, the Rayleigh distribution is a crude approximation. There are other more complicated probability laws (two- or three-parameter models) that yield better results. Rayleigh distribution can be used to determine the year-round wind speed distribution for wind energy utilization and for the study of heat loss through buildings, air pollution, pedestrian wind, and so on.

Example 2-1. The annual mean wind speed at 10-m height at a given airport is 13 mph. How often would the airport encounter winds higher than 30 mph? How often would the wind be between 30 and 40 mph?

[Solution]. Based on the Rayleigh distribution, the probability that a given wind speed V is exceeded is

$$P_E(V) = 1 - F(V) = \exp\left(-\frac{\pi V^2}{4\overline{V}^2}\right) \qquad (a)$$

When $\overline{V} = 13$ mph, the equation becomes

$$P_E(V) = \exp(-0.00465 V^2)$$

Therefore, the probability of wind speed exceeding 30 mph is $P_E(30) = 0.0153$ or 1.53%, and the probability of wind speed exceeding 40 mph is $P_E(40) = 0.0006$ or 0.06%. The probability of wind speed being between 30 and 40 mph is $P_E(30) - P_E(40) = 0.0153 - 0.0006 = 0.0147 = 1.47\%$.

From the foregoing calculations, in 1.53% time of a year, the wind speed is expected to exceed 30 mph, and in 1.47% time of a year, speed is between 30 and 40 mph.

2.5 PROBABILITY OF EXTREME WINDS

The wind speed used in structural design is the extreme (maximum or fastest) values for a given period such as 50 years. To model extreme wind speeds, one must distinguish thunderstorm winds (i.e., thunderstorm-related nontornadic winds) from hurricane and tornado winds for they follow different probability laws.

Thunderstorm Wind Probability

The extreme wind data for thunderstorm winds can be adequately represented by the following probability distribution function:

$$F(V) = \exp\left[-\exp\left(-\frac{a+V}{b}\right)\right] \quad (2.7)$$

and

$$a = 0.450s - \overline{V}$$
$$b = 0.780s \quad (2.8)$$

where V is the annual fastest wind speeds such as listed for Columbia in Table 2.1, $\overline{V}$ is the mean of V, and s is the standard deviation of V.

It can be proved that for this type of distribution, the wind speed V corresponding to a return period T_R (in years) is

$$V = \overline{V} + 0.78 \, (\ln T_R - 0.577)s \quad (2.9)$$

The probability distribution given by Eq. 2.7 is referred to in the literature under different names such as **Type-I distribution, Gumbel distribution, or Fisher-Tippett distribution.** The ANSI standard on wind load (ANSI/ASCE 1988) uses this distribution to determine the maximum fastest-mile wind speed corresponding to different return periods. For instance, based on the 10-year data for Columbia given in Table 2.1, the mean wind speed $\overline{V}$ and the standard deviation are, respectively, 56.3 mph and 4.47 mph. Substituting these values into Eq. 2.9 yields a 68-mph wind for a 50-year return period. Note that based on a much larger body of data, the wind speed given in the ANSI/ASCE standard for Columbia (see "X" marked on Figure 7.1) is 70 mph. The wind speed contours for the United States shown in Figure 7.1 were determined essentially from the foregoing approach using the extreme wind data of 129 stations that had reliable wind records (Simiu, Changery, and Fillibean 1980).

For stations with very short records, such as a few years, analyses cannot be made based on the annual maximum. An alternative is to use the maximum wind of each months. Based on the monthly maxima, the wind speed corresponding to return period T_R is

$$V = \overline{V}_m + 0.78 \, [\ln(12T_R) - 0.577]s_m \quad (2.10)$$

where $\overline{V}_m$ and s_m are, respectively, the mean and the standard deviation based on monthly maximum values of V, and T_r is again in years.

Sec. 2.5 Probability of Extreme Winds

Equation 2.10 can be used for structural design in special wind regions (mountainous areas) where wind records are only a few years long. Note that the contours in Figure 7.1 do not cover these regions.

Example 2-2. A hotel is to be built in a ski resort in Colorado where there is no long-term wind record. The engineer in charge installed an anemometer at the site and recorded the wind speed for a year. He found that the mean and the standard deviation of the monthly maximum wind speed are 40 mph and 10 mph, respectively. Estimate the design wind speed corresponding to a 50-year return period.

[Solution]. For this case, $\overline{V}_m = 40$ mph, $S_m = 10$ mph, and $T_R = 50$ years. Substituting these values into Eq. 2.10 yields $V = 85.4$ mph. Therefore, the wind speed corresponding to a 50-year return period wind is estimated to be 85.4 mph.

Wind speed correlation between sites. The use of short-term monthly maxima to predict long-term extreme wind speeds as done in the foregoing example is unreliable. It should not be used except as a last resort. A better approach is to correlate the short-term wind data of a site to the data of a neighboring site that has a long record. For instance, if in the previous example a 40-year record exists at an airport 80 miles from the ski resort, this long record can be utilized to predict the 50-year return-period wind at the ski resort in the following manner:

Using the Type-I distribution for the 40-year record, suppose the 50-year return-period wind for the airport is calculated to be 70 mph. Furthermore, suppose the mean monthly maximum wind speed at the airport for the same year as measured at the ski resort is 32 mph. The ratio of the corresponding monthly maxima for the ski resort and the airport is $40/32 = 1.25$. Using this same ratio for long-term records, the 50-year extreme wind at the ski resort is estimated to be $70 \times 1.25 = 87.5$ mph.

The foregoing calculation can be generalized. Suppose the extreme wind speeds corresponding to a return period T are $V_1(T)$ and $V_2(T)$ for two neighboring sites 1 and 2, respectively, and suppose the short-term means (averages) for the two sites are U_1 and U_2. Then,

$$V_1(T) = \frac{V_2(T) \, U_1}{U_2} \quad (2.11)$$

Equation 2.11 can be used to calculate the extreme wind speed of any site using the long-term record that exists at a neighboring site. The method is accurate only if the two sites are near each other so that they are subject to the same types of storms affecting the region. Note that the short-term mean wind speeds U_1 and U_2 need not be the mean of the monthly maxima. They can be the mean annual wind speeds of a given year for two neighboring sites. Due to seasonal variation of winds, U_1 and U_2 must be based on a minimum record of 12 months.

Hurricane Wind Probability

The Type-I distribution has been used widely for extreme wind analysis not only in nonhurricane regions but also in hurricane regions. However, it should be realized that when analyzing the annual fastest-wind data for any site in a hurricane region, the data are a mixture of hurricane winds and thunderstorm winds. They cannot be modeled properly by a single probability distribution function, no matter what probability law is used. To model the two separately is often not feasible because hurricanes do not strike the same location each year. There may be insufficient data for hurricane at a given station, and besides, the fastest hurricane wind speed for most years may be zero. To model the hurricane wind speed properly would require a much longer record than most stations have.

For stations that have a relatively long (over-50-year) record of hurricane wind speeds, one can list the highest speed of each hurricane measured at a station in a way similar to Table 2.1, except that some years may have more than one values listed whereas some other years may have none. Such a listing is called a **partial duration series,** which is different from the **annual series** listed in Table 2.1. The latter should not be used for hurricanes.

Suppose r is the rank (order) of a hurricane wind speed in a partial duration series containing M hurricanes in a record of N years. The average number of times per year that this hurricane wind speed will be exceeded is

$$m = \frac{r}{N} \tag{2.12}$$

The reciprocal of m is the average number of years between hurricanes with wind speed greater than that of the rth ranked hurricane. While P_E in Eq. 2.1 is probability, m in Eq. 2.12 is not probability. The value of m can be greater than 1.0 if on the average more than one hurricane per year pass through the station, as it is the case in some typhoon regions of the world such as Taiwan.

Sec. 2.5 Probability of Extreme Winds

TABLE 2.2 Relation Between T_R and T_R'

T_R (Annual series)	1.1	1.5	2	5	10	25	50	100	500
T_R' (Partial duration series)	0.417	0.91	1.44	4.48	9.49	24.5	49.5	99.5	499.5

The exceedance probability based on the partial duration series can be calculated from

$$P_E = \frac{r}{nN} \quad (2.13)$$

where n is the average number of hurricanes per year for the record, namely, $n = M/N$.

To use partial duration series data to determine the relationship between V and T_R, Eq. 2.9 is rewritten as

$$V = \overline{V}' + 0.78 \,(\ln T_R' - 0.577)s' \quad (2.14)$$

where the primed quantities are values calculated from the partial duration series. The quantity T_R' is related to T_R as follows:

$$T_R' = -\frac{1}{\ln\,[1 - (1/T_R)]} \quad (2.15)$$

Values of T_R' calculated from Eq. 2.15 are given in Table 2.2. Note that the values of T_R' approach that of T_R as the return periods get large.

Eqs. 2.14 and 2.15 can be used simultaneously to determine the relationship between V and T_R. It is illustrated in the following example.

Example 2-3. Suppose that a 20-year record exists for a given locality affected by hurricanes. The record contains the maximum wind speeds of eight hurricanes, the average of which is $\overline{V}' = 32$ m/s and the standard deviation is $s' = 12$ m/s. Determine the wind speed corresponding to a 50-year return period.

[Solution]. From Eq. 2.14,

$$\begin{aligned}V &= 32 + 0.78\,(\ln T_R' - 0.577) \times 12 \\ &= 26.6 + 9.36 \ln T_R'\end{aligned} \quad \text{(a)}$$

From Table 2.2, $T_R' = 49.5$ when $T_R = 50$. Therefore, Eq. (a) yields 63.1 m/s.

The variation of hurricane wind speeds along the Gulf and Atlantic coastlines of the United States has been determined by Batts, Russell, and Simiu (1980) using a complicated random simulation technique called the **Monte Carlo simulation.** The result is very useful, and it has been used in ANSI Standard A58.1-1982 and the subsequent ANSI/ASCE 7-88 to specify the values of wind speeds along the hurricane coastline of the United States; see Figure 7.1. Readers interested in this Monte Carlo simulation technique should read Russell (1971); Batts, Russell, and Simiu (1980); and Simiu and Scalan (1986).

Tornado Wind Probability

Due to the lack of tornado wind speed records, tornado wind speed probability can only be estimated. There are several kinds of tornado probability. The **strike probability** P_1 is the probability that a tornado will strike a given point or location within a given year. The **speed probability** P_2 is the probability that a tornado will have a wind speed greater than a certain value V. The probability P that a tornado with

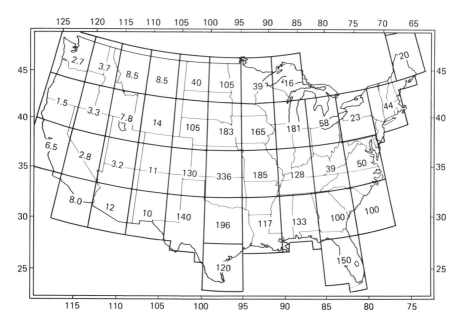

Figure 2.2 Tornado strikes probability, P_i, for various regions of the United States. Units are 10^{-5} probability per year, and each number represents a 5-degree latitude-longitude square. (From Markee, Beckerley, and Sanders, 1974.)

Sec. 2.5   Probability of Extreme Winds

wind speed greater than V will strike a given location in a year is the product of P_1 and P_2, namely,

$$P(V) = P_1 P_2 \qquad (2.16)$$

The strike probability P_1 is calculated from

$$P_1 = n' \frac{A_d}{A_o} \qquad (2.17)$$

where A_d is the average damage area of a tornado, A_o is a reference area such as 100 square miles or a 1-degree longitude-latitude square, and n' is the average number of tornadoes per year in the area A_o.

Figure 2.2 gives the values of P_1 for different regions of the United States. The speed probability P_2 is estimated from a large number of damage surveys using the Fujita Scale (Table 1.3) to determine the wind speed of each tornado. Figure 2.3 gives the values of P_2 based on the average condition of tornadoes occurring anywhere in the nation. From Figures 2.2 and 2.3 and Eq. 2.16, the probability of having a tornado

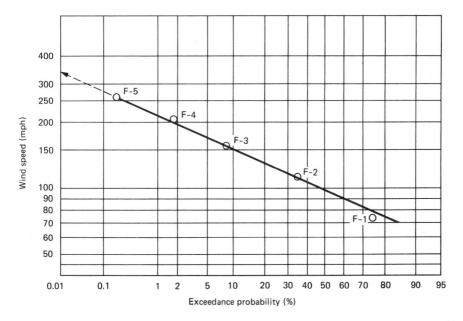

Figure 2.3   Tornado wind speed probability P_2. (Same as the fraction of tornadoes having a wind speed higher than V. Note that the wind speed used in this graph is gust speed). (From Markee, Beckerley, and Sanders, 1974).

with a wind speed greater than 150 mph at a particular location in the state of Kansas during a given year is

$$P(V > 150 \text{ mph}) = P_1 \times P_2 = 336 \times 10^{-5} \times 0.1 = 3.4 \times 10^{-4}$$

Note that the wind speed used in calculating tornado probability is the fastest-gust speed rather than the fastest-mile wind. From Figure 2.3, more than 50% of tornadoes have maximum gust speeds lower than 100 mph, and only 10% of tornadoes have maximum speeds higher than 150 mph. Because the wind speeds used in deriving Figure 2.3 are estimated rather than measured values, any wind speed predicted from this graph can easily be in error by 20%.

The exceedance probability of tornado winds may be compared to that of thunderstorm winds as shown in Figure 2.4 for Kansas City, Missouri. Since the tornado wind speed is based on gust speed, whereas

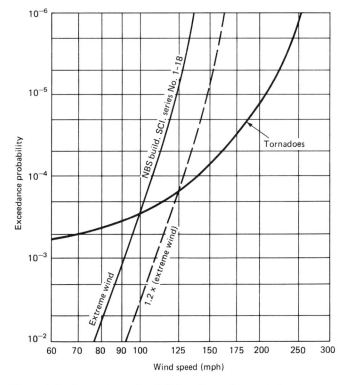

Figure 2.4 Comparison of probability of tornadoes with thunderstorm winds for Kansas City, Missouri. (From Darkow 1986.)

the thunderstorm wind speed is based on the fastest mile, the thunderstorm speed has been converted to gust speed by using a multiplication factor of 1.2. The result is shown by the broken line in Figure 2.4. Comparison of this broken line with the line for tornadoes shows that for wind speeds less than 125 mph the probability of thunderstorm wind is greater than the probability of tornadoes, and the opposite holds for wind speeds greater than 125 mph. This explains why straight-line winds cause far more cumulative damage than tornadoes, and why it is permissible to neglect tornadoes in wind load determination when the design is based on 25-, 50-, or even 100-year return-period winds as in common practice.

Example 2-4. The Nuclear Regulatory Commission (NRC) requires that all nuclear power plants be designed against tornadoes of exceedance probability of 10^{-7}. Determine the corresponding wind speed for power plants in the state of New York.

[Solution]. From Eq. 2.16, $P = P_1 P_2 = 10^{-7}$. From Figure 2.2, for New York $P_1 = 23 \times 10^{-5}$. Therefore, $P_2 = 10^{-7}/(23 \times 10^{-5}) = 4.35 \times 10^{-4} = 0.0435\%$. From Figure 2.3, the corresponding design wind speed is approximately 310 mph. For simplicity, NRC requires all nuclear plants in Region I, including New York, be designed for maximum wind speeds of 360 mph; see Figure 1.9 and Table 1.4.

2.6 SIGNIFICANCE OF RETURN PERIOD AND EXCEEDANCE PROBABILITY

As an illustration, consider a 50-year wind in Columbia, Missouri. From ANSI standard (see Figure 7.1), this corresponds to a fastest-mile wind of 70 mph. It means on the average, Columbia has experienced and is expected to experience a fastest-mile wind *greater than* 70 mph on the average of once in 50 years.

A return period of 50 years corresponds to a probability of $1/50 = 0.02 = 2\%$ for occurrence within a given year. Thus, the probability that a wind exceeding 70 mph will occur within a given year in Columbia is 2%. Suppose a building having a service life of 50 years is designed in Columbia using a basic wind speed of 70 mph corresponding to a 50-year return period. What is the probability that this speed will be exceeded within the lifetime of the structure?

The probability that this wind speed will *not* be exceeded in any year is $1 - (1/50) = 49/50$. The probability that this speed will *not* be exceeded in 50 years in a row is

$$\frac{49}{50} \times \frac{49}{50} \times \cdots \frac{49}{50} = \left(\frac{49}{50}\right)^{50}$$

Therefore, the probability that this wind speed will be exceeded at least once in 50 years is

$$1 - \left(\frac{49}{50}\right)^{50} = 1 - 0.364 = 0.636 \approx 64\%$$

The foregoing calculation shows that the use of 50-year wind to design structures results in a high probability of the design wind being exceeded within the lifetime of the structures. This, however, need not worry the reader since having a wind higher than the design wind does not necessarily mean the building will collapse or even suffer damage. Due to the use of required safety factors such as load factors in structural design, a structure is usually able to resist wind speeds higher than the design wind. Furthermore, designs are made normally by assuming that the wind is perpendicular to the building side having maximum surface area—the worst case possible. In reality, even when a building is hit by a wind higher than the design value, the wind may not be from the worst direction. Therefore, the probability that a building designed using a 50-year return-period wind will be destroyed or seriously damaged by wind within 50 years is much less than 64%.

The foregoing calculations can be generalized as follows:

The probability that any wind speed V will *not* be exceeded in any year is

$$P \text{ (not exceeded in a year)} = 1 - P_E = 1 - \frac{1}{T_R} \quad (2.18)$$

where P_E and T_R are, respectively, the exceedance probability and the return period corresponding to V.

The probability of V *not* being exceeded N years in a row is

$$P \text{ (not exceeded in } N \text{ successive years)} = (1 - P_E)^N \quad (2.19)$$

The probability that V will be exceeded at least once in N years is

$$P \text{ (exceeded within } N \text{ years)} = 1 - (1 - P_E)^N \quad (2.20)$$

Using Eq. 2.20, the probability that a structure will experience greater than 100-year winds (1% exceedance probability) during its lifetime of 50 years is

$$P = 1 - (1 - 0.01)^{50} \approx 40\%$$

In contrast, the probability that a structure under construction (over a period of 6 months) will encounter a 10-year wind (0.1 annual probability) is only

$$P = 1 - (1 - 0.1)^{0.5} \approx 5\%$$

This gives justification for using a much smaller design wind speed and return period for structures under construction than for completed structures.

More about the probability of extreme winds can be found in Simiu and Scanlan (1986).

REFERENCES

ANSI/ASCE 7 (1988). *Minimum Design Loads for Buildings and Other Structures,* American Society of Civil Engineers, New York.

BATTS, M. E., RUSSELL, L. R. AND SIMIU, E. (1980). "Hurricane Wind Speeds in the United States," *Journal of Structural Division,* ASCE, 100(10), 2001–2015.

DARKOW, G. L. (1986). "Tornado Wind Probabilities for Engineers," Course Notes, 11th Annual Continuing Education Short Course on Wind Effects on Buildings and Structures, Engineering Extension, University of Missouri-Columbia.

MARKEE, E. H., BECKERLEY, J. G. AND SANDERS, K. E. (1974). *Technical Basis for Interim Regional Tornado Criteria,* WASH-1300 (UC-11), U.S. Atomic Energy Commission, Office of Regulation, Washington, D.C.

RUSSELL, L. R. (1971). "Probability Distribution for Hurricane Effects," *Journal of Waterways, Harbour and Coastal Engineering Division,* ASCE, 97(2), 193–154.

SIMIU, E., CHANGERY, M. J. AND FILLIBEN, J. J. (1980). "Extreme Wind Speeds at 129 Airport Stations," *Journal of the Structural Division,* ASCE, 106(4), 809–817.

SIMIU, E. AND SCANLAN, R. H. (1986). *Wind Effects on Structures* (2nd ed.), John Wiley & Sons, New York.

3

Wind Characteristics

Since ordinary structures are placed on the ground, for structural design purpose it is important to deal only with **surface winds**, that is, winds near the ground surface. The ensuing discussion of wind characteristics will focus on surface winds. Unless otherwise specified, we shall consider surface winds as the winds at 10-m height above ground.

3.1 VARIATION OF WIND VELOCITY WITH HEIGHT AND ROUGHNESS

Variation of Wind Speed with Height

An important characteristic of wind is the variation of speed with height. The local mean velocity, hereafter referred to simply as **wind speed,** is zero at the surface, and it increases with height above ground in a layer within approximately 1 kilometer from ground called the **atmospheric boundary layer**. Above this layer of thickness δ exists the **gradient wind,** which does not vary with height; see Figure 3.1. The wind speed profile within the atmospheric boundary layer belongs to the turbulent boundary layer type, which can be approximated either by a logarithmic equation or a simpler power-law formula as follows:

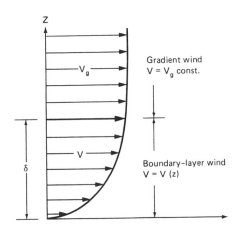

Figure 3.1 Wind velocity profile in atmospheric boundary layer. (Note the existence of gradient wind with constant velocity above the boundary layer).

Logarithmic law

$$V(z) = \frac{1}{\kappa} V_* \ln \frac{z}{z_0} \quad (3.1)$$

Power law

$$V(z) = V_1 \left(\frac{z}{z_1}\right)^\alpha \quad (3.2)$$

where $V(z)$ is the velocity (wind speed) at height z above ground; V_* is the **shear velocity** or **friction velocity**, which by definition is equal to $\sqrt{\tau_0/\rho}$, where τ_0 is the stress of wind at ground level and ρ is the air density; κ is the **von Karman constant** equal to 0.4 approximately; z_0 is the **roughness** of ground, which is an effective height of ground roughness elements; V_1 is the wind speed at any reference height z_1; and α is the **power-law exponent,** which depends on roughness and other conditions.

Note that both Eqs. 3.1 and 3.2 are empirical formulas for wind speed variation over flat (horizontal) areas. Although the logarithmic law is slightly more accurate for large heights, it is somewhat more difficult to use, and it produces an unreal negative speed at heights $z < z_0$. For these reasons, in engineering applications and in building codes/standards, the power law is used most often.

Variation of Wind Speed with Surface Roughness

Surface roughness has a profound effect on wind speed. The rougher a terrain is, the more it retards the wind in the atmospheric boundary layer. The retardation of wind by increased roughness is manifested by an increase in the value of z_0 in the logarithmic law. The way surface roughness affects the power law is considered next in detail.

Since the power law is valid for any value of z within the atmospheric boundary layer δ, we can set $V_o = V_g$ at $z = \delta$, which results in

$$V = V_g \left(\frac{z}{\delta}\right)^\alpha \tag{3.3}$$

Surface roughness affects Eq. 3.3 through the values of α and δ. In general, the rougher a surface is, the higher the values of α and δ, and the smaller the velocity V is at any given height z.

If winds were quasi-steady (i.e., having mean values constant with time), the power-law exponent α would vary from about 1/7 for smooth surfaces to about 1/2 for rough terrains. However, since winds are seldom quasi-steady, the range of α is normally between 1/10 and 1/3. A former U.S. standard on wind load ANSI A58.1-1972 which is based on the fastest-mile wind uses an α value ranging from 1/7 for the smoothest terrains to 1/3 for the roughest terrains. The 1982 edition of ANSI and the 1988 edition of ANSI/ASCE-7 (see Chapter 7) have changed the range to 1/10–1/3 (see Table 3.1). In the contemporary Australian standard, which is based on 2- to 3-second gusts, the range of α is 0.07–0.2. As will be shown later, the 2- to 3-second gust is a much shorter averaging time than is the fastest-mile wind. A shorter averaging time

TABLE 3.1 Values of z_0, D_o, α, and δ for Different Exposure Categories (Terrain Conditions) Used in ANSI A58.1-1982 and ANSI/ASCE-7-1988

Exposure Category	Terrain Roughness z_0 (cm)	Surface Drag Coefficient D_o	Power-Law Exponent α	Atmospheric Boundary-Layer Thickness δ	
				(ft)	(m)
A	80	0.0251	1/3	1500	457
B	20	0.0105	2/9	1200	366
C	3.5	0.0050	1/7	900	274
D	0.7	0.0030	1/10	700	213

Note: A = large cities, B = urban and suburb, C = open terrain, D = open coast.

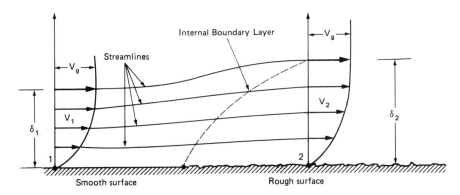

Figure 3.2 Change of wind-speed distribution with terrain roughness. (The development of the internal boundary layer at the place of roughness change.)

justifies the use of a smaller α because it takes time for a boundary layer to develop. The variation of wind speed with height for peak gusts which have a short averaging time is not as drastic as for peak winds of longer durations such as the fastest mile.

Equation 3.3 is used in ANSI A58.1-1982 and ANSI/ASCE-7-1988 to adjust wind speed where the terrain condition (exposure category) changes, as can be illustrated as follows:

Consider the change of terrain from smooth to rough as shown in Figure 3.2. Due to the greater roughness at location 2 than at location 1, winds must slow down from 1 to 2, resulting in a rise in streamlines as shown in the graph. This expansion of streamlines causes an increase in the boundary-layer thickness (gradient height) from δ_1 to δ_2 and an increase in the power-law exponent from α_1 to α_2. The gradient wind velocity, V_g, however, remains unchanged. From Eq. 3.3,

$$V_1 = V_g \left(\frac{z_1}{\delta_1}\right)^{\alpha_1} \tag{3.4}$$

and

$$V_2 = V_g \left(\frac{z_2}{\delta_2}\right)^{\alpha_2} \tag{3.5}$$

Dividing Eq. 3.5 by Eq. 3.4 yields

$$\frac{V_2}{V_1} = \left(\frac{z_2}{\delta_2}\right)^{\alpha_2} \left(\frac{\delta_1}{z_1}\right)^{\alpha_1} \tag{3.6}$$

Sec. 3.1   Variation of Wind Velocity With Height and Roughness   45

Example 3-1. Consider a sudden change of terrain roughness from $\alpha_1 = 1/7$ to $\alpha_2 = 1/3$ and a corresponding change of gradient height from $\delta_1 = 274$ m to $\delta_2 = 457$ m (see Table 3.1). Suppose $V_1 = 50$ m/s at $z_1 = 20$ m. What is the wind speed at 20 m above ground at location 2 (i.e., at $z_2 = 20$ m)?

{Solution} The wind speed at 20 m above ground at location 2 is, from Eq. 3.6,

$$V_2 = 50 \times \left(\frac{20}{457}\right)^{1/3} \times \left(\frac{274}{20}\right)^{1/7} = 25.6 \text{ m/s}$$

This example shows that by changing from a smooth terrain (category C) to a very rough terrain (category A), the wind speed at the same height (20 m in this case) is almost reduced to half its original value.

Returning to Eq. 3.1, if $V = V_1$ at $z = z_1$, we have

$$V_* = \frac{0.4 \, V_1}{\ln(z_1/z_0)} \qquad (3.7)$$

Equation 3.7 can be used for calculating V_* from known values of V_1, z_1, and z_0. If z_0 is not known, the equation can be used to calculate both V_* and z_0 if the values of velocity are known at two different heights.

The wind-generated shear stress at the surface of ground is

$$\tau_o = D_o \rho V_1^2 \qquad (3.8)$$

where V_1 is the wind speed at a reference height, normally taken at 10 m above ground, and D_O is the **surface drag coefficient.**

From Eq. 3.8 and the definition of V_*,

$$D_o = \frac{V_*^2}{V_1^2} \qquad (3.9)$$

Substituting Eq. 3.7 into 3.9 for $z_1 = 10$ m yields

$$D_o = \left[\frac{0.4}{\ln(10/z_o)}\right]^2 \qquad (3.10)$$

Eq. 3.10 can be used to calculate D_o from roughness height z_o as listed in Table 3.1. Note that the value of z_o in Eq. 3.10 must be in meters rather than in any other units.

Variation of Wind Direction with Height

Due to ground resistance, the Coriolis force causes winds to change direction continuously with height, resulting in a spiraling of the wind vector with height called the **Ekman spiral**. Because only small changes of wind direction occur within a few hundred meters above ground, the change of wind direction with height can be neglected safely in structural design. It should be considered, however, in the study of long-distance transport of pollutants such as those generated from power-plant stack emission.

3.2 TURBULENT CHARACTERISTICS OF WIND

Averaging Time

Wind is a turbulent flow, characterized by the random fluctuations of velocity and pressure. If the instantaneous velocity of wind at a given point is recorded as a function of time on a chart, the result will look like that in Figure 3.3.

Due to its fluctuating nature, the characteristics of wind velocity are to be studied statistically. An important statistical property is the

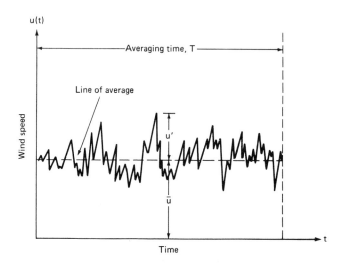

Figure 3.3 Typical trace of variation of wind speed u with time t.

local mean speed which is the wind speed at a given location averaged over a certain duration. In the literature, the local mean is sometimes referred to by other names such as **temporal mean** or **time-averaged** values. The velocity (wind speed) V discussed in the previous section is the local mean speed. Because wind speed changes constantly, different averages are obtained by using different averaging times or durations. For instance, while a 5-minute average of the peak of a high wind may yield 60 mph, the same peak averaged over an hour may be only 40 mph. This shows that when mentioning the mean or average speed of wind, one must specify the averaging time.

The longest averaging time used for wind speed is the operational period of the measuring station, say, 50 years. This long-term average is often referred to as the **annual mean** or the **long-term average**. For instance, the annual mean wind speed for Kansas City is only about 11 mph. Although information on this speed is important for wind energy utilization, it is useless for wind load on structures because only high winds of short durations are of interest in this case. The wind speed used in structural design is the peak value averaged over a given period. The longest averaging time for peak values used in structural design is an hour—the Canadians use hourly extreme wind for structural design. The shortest is 2 to 3 seconds—the gust speed measured by ordinary anemometers. Both the British and the Australian standards on wind load are based on such a gust speed.

In general, as the averaging time decreases, the peak wind speed for a given return period increases. Suppose that V_T is the peak wind speed based on an averaging time of T seconds and V_H is the peak based on the hourly average. The relationship between V_T and V_H for smooth, open terrains is given approximately in Figure 3.4. From this graph, the 2-second gust speed above open terrains is about 1.53 times the speed of the peak hourly wind.

Fastest-Mile Wind

The wind speed used in both the 1972 and 1982 editions of ANSI A58.1 is the **fastest-mile wind** V_F which is the peak wind speed averaged over 1 mile of wind passing through the anemometer. The averaging time of the fatest-mile wind is

$$T = \frac{3600}{V_F} \qquad (3.11)$$

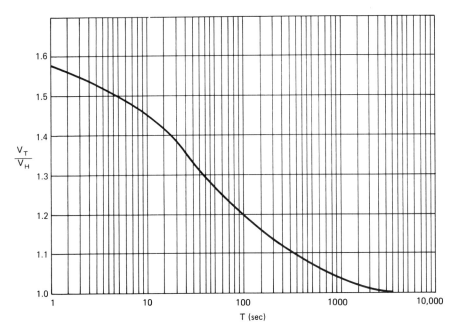

Figure 3.4 Variation of peak wind speed with averaging time. (V_t = peak wind speed averaged over time T; V_H = peak wind speed averaged over an hour.)

where T is in seconds and V_F is in mph. If $V_F = 60$ mph, the averaging time will be $T = 3600/60 = 60$ s. If V_F is 120 mph, T will be decreased to 30 seconds. From the foregoing, the averaging time of the fastest-mile wind used in the design of structures, which normally use a wind speed between 60 and 120 mph, is from 30 to 60 seconds.

Gust Factors

Gust is the rapid fluctuations or instantaneous velocity of wind. Ordinary structures are sensitive to peak gusts of a duration of the order of 1 second. Therefore, the use of any mean wind speed (such as the fastest-mile wind) that has a much longer duration than 1 second without taking into account gust effect is inadequate for structural design. One must design structures to withstand gusts rather than the mean wind speed. The gust speed, V_G, can be determined from the mean wind speed, V, by using the following equation,

$$V_G = G_t V \tag{3.12}$$

Sec. 3.2 Turbulent Characteristic of Wind

where G_v is called the **velocity gust factor**. Another gust factor, the **pressure gust factor** G_p, can be used to relate the pressure p generated by the mean wind, to the pressure generated by gust, p_G, in the equation.

$$p_G = G_p p \qquad (3.13)$$

Since p is proportional to V^2, it follows that $G_p = G_v^2$.

If we assume that the gust speed is based on 1-second average and the fastest-mile wind is based on 30-second average (i.e., $V = 120$ mph), from Figure 3.4 the gust factors are

$$G_v = \frac{1.56}{1.32} \approx 1.2 \quad \text{and} \quad G_p \approx 1.2^2 \approx 1.4$$

Because Figure 3.4 is for smooth terrains, in the case of rough terrains the gust factors G_v and G_p should be greater than those just calculated.

Not all buildings and structures are equally sensitive to gust. In general, the more flexible a structure is, the more sensitive it is to gust. Small structures are also more sensitive to gust than large structures. Therefore, in the design of structures without having to resort to a dynamic analysis, the pressure gust factor used must be different from the value of G_p given in Eq. 3.13. In the United States, the gust factor used in structural design is called the **gust response factor**, G, which is based on the fastest-mile wind speed. In Canada, there the mean wind speed is based on the fastest hourly average, it is called the **gust effect factor**, see Chapter 7. The values of the gust effect factor is higher than that of the gust response factor because the fastest-mile wind speed is higher than the fastest hourly wind. Note that both the gust response factor and the gust effect factor are functions of G_p and the characteristics of structures. More about G will be discussed in Chapter 5.

Turbulence Intensity

Turbulence is the fluctuating velocity component of flow. The wind near ground level is highly turbulent. Its velocity vector $\vec{V}$ at any time t can be decomposed into three components u, v, and w, respectively in the longitudinal (horizontal), vertical, and lateral directions. Each component can further be decomposed into a mean (temporal average) and a fluctuating component as follows,

$$u = \bar{u} + u'$$
$$v = \bar{v} + v'$$
$$w = \bar{w} + w' \quad (3.14)$$

where the bar over any quantity marks the temporal average and the prime denotes fluctuating components. The three terms of $u = \bar{u} + u'$ are illustrated in Figure 3.3.

Note that turbulence is always three dimensional even if the mean velocity of flow is one or two dimensional. For instance, although the wind over a large flat area is essentially horizontal ($\bar{u} = V$, $\bar{v} = 0$, and $\bar{w} = 0$), all the three components of turbulence u', v', and w' exist. Because u' is the strongest turbulence component most relevant to structural design, the ensuing discussion of turbulence intensity, spectrum, and so on will deal with this component only. Readers can extend the same concept discussed to v' and w'.

A measure of the intensity of turbulence is the root-mean-square (rms) value of u', namely,

$$u_I \text{ (turbulence intensity)} = \sqrt{\overline{u'^2}} \quad (3.15)$$

where the bar inside the square-root sign represents temporal mean.

The value of turbulence intensity u_I divided by the mean velocity $\bar{u}$ is called the **relative intensity of turbulence** or the **turbulence level**, I_r. In the literature on wind load, the term "turbulence intensity" is sometimes used to represent the relative intensity of turbulence, which is misleading. Surface winds at 10-m height often have relative turbulence intensity of the order of 0.2 (20%) or 0.3 (30%). The relative intensity increases with ground roughness and decreases with height. It also varies with the duration (averaging time) used in determining the mean velocity $\bar{u}$—longer durations yield smaller mean velocities and hence larger values of relative intensity. Ordinarily, as the mean wind speed at a given location increases, the turbulence intensity at the same location also increases in proportion, causing the relative intensity to remain constant. An approximate formula to estimate the magnitude of the relative intensity of turbulence is

$$I_r \text{(relative intensity)} = \frac{u_I(z)}{V_H(z)} = 0.1 \left(\frac{\delta}{z}\right)^\alpha \quad (3.16)$$

where V_H is the mean hourly wind speed and δ is the boundary layer thickness. The values of δ and α can be obtained from Table 3.1. Eq. 3.16 should not be used for heights z less than 5 m.

Sec. 3.2 Turbulent Characteristic of Wind   51

Example 3-2. The fastest-mile wind speed at the roof level of a building 100 m high is 80 mph. The terrain is rough (Exposure Category B in Table 3.1). What is the turbulence intensity of the wind encountered by the roof?

[Solution] From Eq. 3.11, the averaging time for the 80-mph fastest-mile wind is $T = 3600/80 = 45$ seconds. Then, from Figure 3.4., $V_T/V_H = 1.28$ or $V_H = 80/1.28 = 63$ mph. From Table 3.1 with Exposure Category B, $\alpha = 2/9$ and $\delta = 366$ m, and from Eq. 3.16, $u_I/V_H = 0.133$. Therefore, the turbulence intensity (rms value) at the roof level is $u_I = 0.133 V_H = 8.4$ mph.

Spectrum of Turbulence

The turbulence of wind consists of a large number of eddies or velocity waves having different amplitudes and frequencies. The same concept of the spectrum of light applies to turbulence.

Consider that the turbulence intensity (i.e., the rms of u') for wind eddies in the frequency range n and $n + dn$ is $S_1 dn$ where S_1 is a function of n. The total intensity of turbulence generated by all the eddies of various frequencies is

$$u_I = \sqrt{\overline{u'^2}} = \int_0^\infty S_1(n)\, dn \qquad (3.17)$$

An alternate type of wind velocity spectrum is S_2 for which the product $S_2\, dn$ gives the mean-square values instead of the root-mean-square values of u', In terms of S_2,

$$u_I^2 = \overline{u'^2} = \int_0^\infty S_2(n)\, dn \qquad (3.18)$$

S_2 is often referred to as the **power spectrum** because it is proportional to the power (kinetic energy per unit time) of the turbulence. In the literature, the power spectrum of any function of time, $X(t)$, is often called the **spectral density function** of X.

The spectrum of wind turbulence, either S_1 or S_2, can be measured by processing the electronic signal (voltage or current output) of an anemometer. The signal is first rid of its mean value—the d.c. (direct current) component. Only the fluctuating part—the a.c. (alternative current) component—is kept and analyzed for determining the spec-

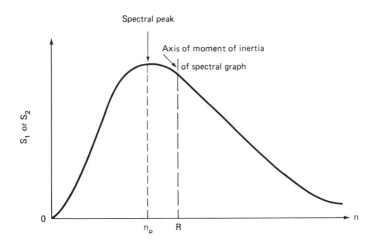

Figure 3.5 Typical spectrum of wind turbulence.

trum. The spectrum can be obtained with a spectrometer which connects the signal (the a.c. component) to a set of electronic bandpass filters each having a narrow bandwidth and a different filter frequency. The signal that has passed through each filter is measured by an rms-meter for its intensity. Dividing each rms signal measured by the band width of the filter yields the value of the spectrum S_1 at the filter frequency. Finally, the spectrometer plots the spectrum, either S_1 or S_2, as a function of filter frequency n. The result is a spectrum as illustrated in Figure 3.5. An alternate method to obtain spectra is to perform a Fourier analysis of signals on a digital computer. Most modern computers, including personal computers, can perform such an analysis and plot the resulting spectra using special software.

The spectrum in Figure 3.5 illustrates certain special features of all turbulence spectra: Due to the d.c. cutoff in turbulence measurements, all spectra pass through the origin of their coordinates. They all increase rapidly to reach a peak at a certain frequency n_p corresponding to the frequency of the strongest eddies in the wind. After passing the peak, spectral curves decrease gradually, approaching zero value when n approaches infinity. The spectral peak n_p for the turbulence in natural winds is usually at rather low frequencies—of the order of 1 cycle per minute. In wind tunnel, n_p is usually of the order of 1 hertz (cycles per second) or higher. Generally, spectral-peak frequency increases as the wind speed increases and vice versa.

Sec. 3.2 Turbulent Characteristic of Wind

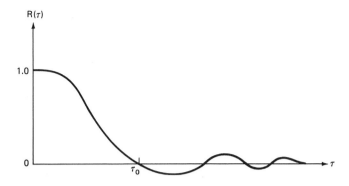

Figure 3.6 Typical variation of autocorrelation $R(\tau)$ with time shift τ for wind turbulence.

Correlation Coefficient of Turbulence

By definition, the correlation coefficient C of two time-dependent variables $X_1(t)$ and $X_2(t)$ is

$$C = \frac{\overline{X'_1 X'_2}}{\overline{X'^2_1}\ \overline{X'^2_2}} \qquad (3.19)$$

where the primes denote fluctuating components (e.g., $x'_1 = x_1 - \bar{x}_1$) and the bar above any quantity denotes the temporal average of the quantity.

Note that the value of any correlation coefficient is between 1 and -1. While $C = 1$ indicates a perfect positive correlation, $C = -1$ indicates a perfect negative correlation. A positive correlation is obtained when two signals are in phase, and a negative correlation results from two signals out of phase. $C = 0$ indicates a complete lack of correlation between two signals.

A special correlation coefficient relevant to wind load determination is the **autocorrelation coefficient** of the longitudinal component of turbulence u', defined as

$$R(\tau) = \frac{\overline{u'(t)\,u'(t + \tau)}}{\overline{u'^2}} \qquad (3.20)$$

where τ is a time shift of the signal u'. The autocorrelation coefficient $R(\tau)$ gives the correlation of the turbulence u' with itself at a time τ later. The typical shape of $R(\tau)$ is shown in Figure 3.6. Note that $R(\tau)$ may oscillate about the τ axis after a time τ_o corresponding to the time that R has first reached zero.

Another correlation coefficient used in wind load study is

$$f(\xi) = \frac{\overline{u'(x)\,u'(x+\xi)}}{\overline{u'^2}} \tag{3.21}$$

where ξ is the increase in distance x. The function $f(\xi)$ gives the correlation of u' with itself at a distance $x = \xi$ apart.

If all the eddies in a turbulent flow are carried downstream by a constant mean flow velocity U (**Taylor's hypothesis**), it can be proven that $\partial/\partial t = U\,(\partial/\partial x)$, and $f(\xi) = R(\tau)$. Although Taylor's hypothesis holds only for uniform flow with homogeneous turbulence, it is often used in turbulent shear flow as an approximation, as in the case of boundary-layer wind. However, it should be realized that in shear flow, the velocity, U, at which eddies are transported downstream is not the same as the local mean velocity $\bar{u}$. At small distances from a wall, the eddy transport velocity is normally much greater than the local mean velocity. This is due to the fact that each eddy is transported at a constant velocity corresponding to the mean velocity across the eddy, $U(z_1)$, where z_1 is the eddy height, rather than the local mean velocity, $\bar{u}(z_2)$, where z_2 is the distance from wall at which the turbulence is measured; see Figure 3.7.

Integral Scales of Turbulence

The **integral time scale** of turbulence is, by definition,

$$T_u = \int_0^\infty R(\tau)\,d\tau \tag{3.22}$$

From Eq. 3.22, the integral time scale equals the area under the autocorrelation coefficient curve in Figure 3.7. The integral time scale represents the average period of large eddies in flow.

Another integral scale is the **integral length scale** defined by

$$L_u = \int_0^\infty f(\xi)\,d\xi \tag{3.23}$$

Note that L_u represents the average longitudinal size of large eddies in the flow. For uniform flow with homogeneous turbulence, the two integral scales are related to each other as follows:

$$L_u = \bar{u}T_u \tag{3.24}$$

Eq. 3.24 makes the integral length scale the same as the average wave length of large eddies in flow.

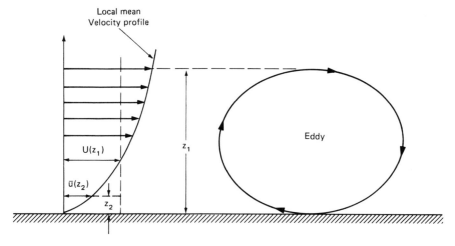

Figure 3.7 Eddy transport velocity $U(z_1)$ compared to local mean velocity $\bar{u}(z_2)$. (Note that $U(z_1)$ is larger than $\bar{u}(z_2)$ when $z_1 \gg z_2$.)

Note that in wind tunnel testing of structures (see Chapter 6), not only must the mean velocity profile and the turbulence intensity in the wind tunnel be similar to those of the natural wind encountered by the prototype, but the spectra of turbulence and the integral scales also must be similar.

3.3 MODIFICATION OF WIND BY TOPOGRAPHY, WOODS, AND STRUCTURES

Topography, woods, buildings, and other structures are all able to alter the direction and the speed of wind. They may either slow down the wind in certain regions—a shielding effect—or accelerate the wind in other regions—an amplifying effect.

Topographical Effect on Wind

Topographical features such as mountains, valleys, hills, canyons, cliffs, and so on can cause drastic change of local wind speed and direction, having a strong effect on structures built on or near these features. The topographical effect on winds in mountain regions is so complex that the ANSI standard A58.1-72&82 and the subsequent standard ANSI/ASCE 7-88 do not attempt to provide any guidelines except to warn designers not to use the wind speed map (Figure 7.1) in mountain regions such as the Rocky Mountains and to seek other solutions such as using local wind data if they exist. One of the topographical effects, the mountain downslope wind, has been discussed in Chapter 1. Other topographical effects on wind are considered next.

Amplification of wind by hills. Ordinarily, the wind speed increases on the windward slope of a hill or a mountain peak, reaching a maximum at or near the summit. Two major factors contribute to this amplification of wind. First, mountains and hills restrict the passage of wind, causing the streamlines of the wind to converge on the windward slope as shown in Figure 3.8. Since wind speed is inversely proportional to the spacing between streamlines of a two-dimensional flow, the wind speed continuously accelerates as the wind approaches a mountain (hill) top. This effect, due to compression of streamlines, is most profound for two-dimensional mountains (hills) with winds perpendicular to ridgelines. This explains why two-dimensional mountains (hills) generally encounter greater amplification of wind than three-dimensional mountains (hills). Second, the height of mountains (hills) brings the gradient wind closer to surface, resulting in a reduction of the gradient height on mountains (hills), which, from Eq. 3.3, causes an increase in the velocity within the boundary layer.

Generally, the wind speed above hills increases rapidly with height in a region very close to the surface. Outside this small surface layer, the increase is at a rate less than for wind above flat surfaces, resulting in a decrease in the α value of the power law given in Eq. 3.3. In some cases, the wind speed above a hill may first increase with height until it reaches a maximum at a certain elevation. Above this height wind speed decreases instead of increases. Such a profile cannot be represented adequately by the power law or the logarithmic velocity distributions, which are for winds above flat areas.

As reported in Chapter 1, the highest typhoon surface wind speeds measured in the world are 85 m/s (189 mph) gust speed at Muroto, Japan, and 75 m/s (167 mph) fastest 10-minute speed at Lan-Yu, Taiwan. The highest surface wind speed ever measured in the world is 231 mph (103 m/s) at Mt. Washington in New Hampshire, United States, April 12, 1934. These record winds were all caused by mountains.

From wind tunnel studies and field measurements, it is known that for a two-dimensional hill with a wind perpendicular to ridgeline, maximum amplification of wind occurs when the windward slope is approximately 1:3.5 (vertical:horizontal) and when the surface of the windward

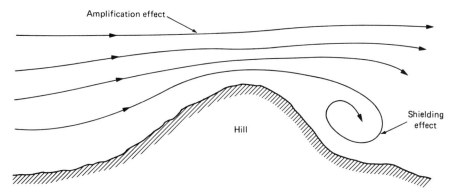

Figure 3.8 Effect of hills on wind field.

slope is smooth. Such a hill can cause surface wind speed to double at hilltop. Since wind load is proportional to the square of wind speed, a structure constructed on the top of such a hill can expect wind loads four times as high as encountered by the same structure on a flat area in the same region. This points to the great importance of treating the orographic effect of wind when designing structures on mountains or hills. As indicated in Fig. 3.8, a hill can provide some shielding of wind on the leeward separation zone and wake.

Amplification factor and speed-up ratio. The amplification of winds by hills, mountains, escarpments, and so on can be quantified by using the **amplification factor,** A, defined as

$$A = \frac{V'(z)}{V(z)} \qquad (3.25)$$

where V' is the amplified wind speed at height z above the surface of a hill or slope and V is the speed of the approaching wind at the same height z above ground. A value of $A = 2$ means the wind speed is doubled by the hill.

The **speed-up ratio or fractional speed-up** ratio β is defined as

$$\beta = \frac{V'(z) - V(z)}{V(z)} \qquad (3.26)$$

From Eqs. 3.25 and 3.26,

$$A = \beta + 1 \qquad (3.27)$$

The range of β is normally between 0 and 1.0; the range of A is between 1.0 and 2.0.

Based on the information reported in Jackson and Hunt (1975), Bowen (1983), Taylor and Lee (1984), and Selvam (1987), a simple method is given here to estimate β (or A) values for three types of topographical features: (1) two-dimensional hills, (2) two-dimensional escarpments, and (3) three-dimensional axisymmetric hills. The method can be used by structural engineers faced with the task of designing structures on hills, ridges, or cliffs.

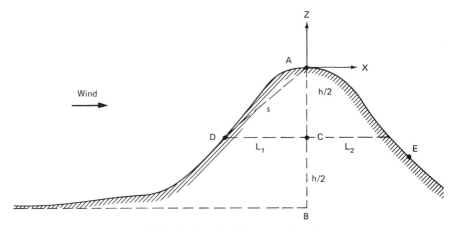

(a) 2 or 3-dimensional hill or mountain peak

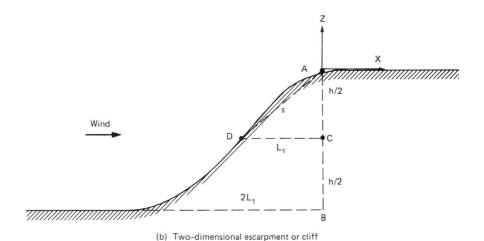

(b) Two-dimensional escarpment or cliff

Figure 3.9 Calculation of speed-up of wind over hills and escarpments.

Sec. 3.3 Modification of Wind by Topography, Woods, and Structures

Consider a two- or three-dimensional hill or mountain peak as shown in Figure 3.9(a). First, locate the summit point A. Then, draw a vertical line through A and a horizontal line along the nearly horizontal ground surface upwind. The two lines intersect at B. The distance AB marks the characteristic height of the hill, h. Locate the midpoint C on line AB. Next, draw a horizontal line through C to intersect the upwind slope at D. The slope s of the straight-line AD is to be used for determining the amplification or speed-up of wind near the summit. A similar procedure should be used to determine the slope s of the top half of an escarpment as shown in Figure 3.9(b). Note that only the top half of the hill or escarpment is used for determining the slope because the top half has more influence on the wind speed near hill tops than has the average slope of the entire windward face of the hill.

Finally, to determine the fractional speed-up ratio β for wind at any horizontal distance x downwind of the summit point and at any vertical distance z above the surface of the hill slope, one can use

$$\beta = bsc \qquad \text{(for mean hourly speeds)} \qquad (3.28)$$

and

$$\beta = \frac{bsc}{1 + 3.7\, I_r} \qquad \text{(for gust speeds)} \qquad (3.29)$$

where b equals 1.6 for escarpment, 4.0 for two-dimensional hills, and 3.2 for three-dimensional hills and I_r is the relative intensity of turbulence, which can be estimated from Eq. 3.16. The values of c can be calculated as follows:
For mild slopes (i.e., $s < 0.3$),

$$c = \left(1 - \frac{|x|}{1.5L}\right) \exp\left(-\frac{2.5z}{L_1}\right) \qquad (3.30)$$

For steep slopes (i.e., $s > 0.3$),

$$c = \left(1 - \frac{|x|}{1.5L}\right) \exp\left(-\frac{1.5z}{h}\right) \qquad (3.31)$$

The values of L in Eqs. 3.30 and 3.31 are determined as follows:
For any point above the windward slope ($x < 0$), $L = L_1$. For any point above the leeward slope ($x > 0$), $L = L_2$ for hills and ridges, and $L = 2L_1$ for escarpments.

Two restrictions must be placed on the use of the foregoing equations:

1. The slope s used in Eq. 3.28 must not be greater than 0.30. Otherwise, use $s = 0.30$.
2. The height z used in Eqs. 3.30 and 3.31 must be greater than $0.05L_1$. Otherwise, use $z = 0.05L_1$.

Example 3-3. Suppose a tower 300 m high is to be built on a 500-m ridge and suppose the windward slope is 0.20. Upwind from the hill at 10-m height above ground, the approaching wind speed is 30 m/s. What are the wind speeds at the location of the tower at 10-m, 30-m, 100-m, and 300-m heights?

[Solution]. With $z = 300$ m, $h = 500$ m, and $s = 0.2$, we have $L_1 = h/2s = 1250$ m, $|x| = 0$, $s < 0.30$, $c = 0.549$ (Eq. 3.30), $b = 4.0$, $\beta = 0.44$ (Eq. 3.28), and $A = 1.44$ (Eq. 3.27). This calculation shows that at $z = 300$ m, the mean hourly wind speed is amplified by a factor of 1.44 at the ridge of this two-dimensional hill. In a similar manner, the mean speed at $z = 100$ m, 30 m, and 10 m are amplified by 1.66, 1.75, and 1.78 times, respectively, as the wind reaches the ridge. Note the greater amplification of the wind at lower heights.

Suppose the upstream wind speed distribution is governed by the power law $\alpha = 1/7$. From Eq. 3.2, the wind speed upstream at any height z is

$$V(z) = 30\left(\frac{z}{10}\right)^{1/7} \qquad \text{(a)}$$

From Eq. (a), the wind speeds upstream at 300-, 100-, 30-, and 10-m heights are, respectively, 48.8, 41.7, 35.1, and 30 m/s. Using the corresponding values of the amplification factor A calculated before, the wind speeds at 300-, 100-, 30-, and 10-m heights above the ridge are, respectively, 70.3, 69.2, 61.4, and 53.4 m/s.

Amplification of winds by canyons. When winds enter a narrow canyon, the streamlines of the wind converge or contract as shown in Figure 3.10, causing an amplification of wind speeds in the narrow part of the canyon. The situation is similar to the increase of fluid velocity at the throat of a Venturi meter. It can be called the **Venturi effect**.

Sec. 3.3   Modification of Wind by Topography, Woods, and Structures   61

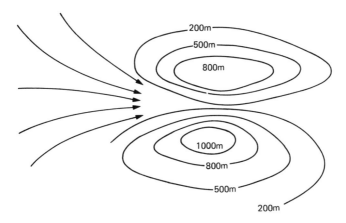

Figure 3.10   Speed-up of wind through canyons.

Amplification of winds by cliffs.   When wind is blowing against a sharp (vertical) cliff of large height, the stagnation pressure generated on the cliff increases with height because the velocity in the approaching wind which produces the stagnation pressure increases with height. This increase of stagnation pressure with height on the windward side of high cliffs causes a strong downdraft near the surface of the cliff and a reversal of wind direction at the bottom of the cliff as shown in Figure 3.11. In addition, amplification of wind occurs both at the windward edge of the cliff (point A in Figure 3.11) and along the edges of the sides (such as line AB) if the cliff has such side edges. The speed-up ratio for cliffs (escarpments) can be determined in the same manner as for hills, as discussed before.

Shielding by topographical features.   Usually, small valleys are protected by the surrounding hills from the onslaught of high winds. Valleys surrounded by high mountains, on the other hand, may be affected by mountain downslope winds.

Mountain and hills that have relatively steep lee slope cause flow separation on the lee side. For instance, if a structure exists at point E in Figure 3.9(a) or point C in Figure 3.11, it would be protected (shielded) by the hill or the cliff. However, since the direction of high winds changes with time, a building that receives shielding from a hill or cliff at one time may encounter amplified wind at another time when the wind direction changes. The change of high wind direction must be considered in the planning and design of buildings near large obstacles such as a hill.

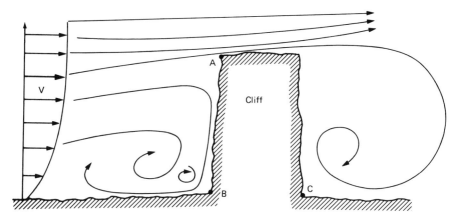

Figure 3.11 Effect of cliff on wind field.

Complex topography. The wind patterns in complex topography such as in mountain regions are so complicated that they defy analysis or accurate description. When determining wind loads in such regions, one must seek local data (long term or short term), conduct field measurements, or perform wind tunnel tests.

Effect of Woods on Wind

Although trees and shrubberies are effective windbreaks, they cannot be counted on to protect structures for they may be destroyed or damaged by high winds before the structures are destroyed or damaged. In fact, in high winds, trees and shrubberies often become a liability to structures. For this reason, normally, one should not reduce the design wind speed in wooded areas. In fact, the designer should consider the probability of structures being hit by falling branches or toppled trees—the wind-generated missiles.

Effect of Building/Structures on Wind

Like topographical features, buildings and other structures deflect winds, causing a change in wind speed and direction around the buildings/structures. This **building-spawned wind** can affect the operation and use of buildings, can cause problems to pedestrians, and can impose large wind loads on adjacent buildings and structures, especially on smaller buildings and structures. A few most noticeable types of building-spawned winds are discussed next.

Sec. 3.3   Modification of Wind by Topography, Woods, and Structures   63

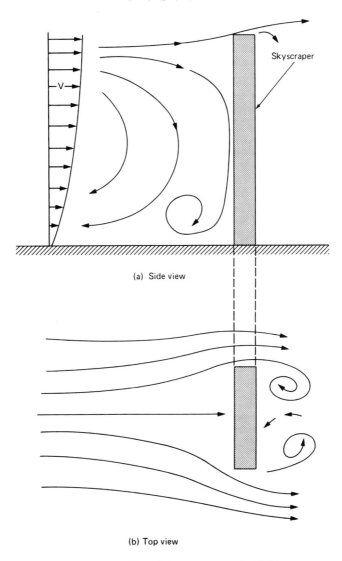

(a) Side view

(b) Top view

Figure 3.12   Effect of skyscrapers on wind field.

Skyscraper-spawned wind. A skyscraper has the same effect as a high cliff. It generates a downdraft along the windward face, a strong ground-level gusty wind upwind of the building and around the two edges of the windward wall, and a turbulent wake behind the building. The situation is depicted in Figure 3.12. An example is the Sears Tower in Chicago—the tallest building in the world in 1990. This building

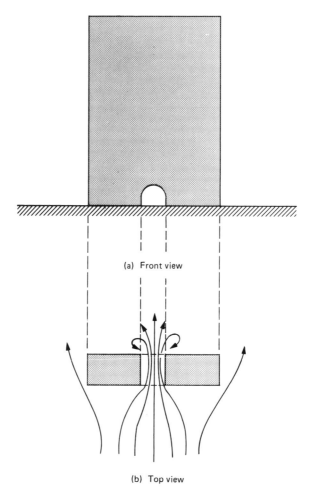

Figure 3.13 Speed-up of wind through arcade.

often causes strong ground-level, gusty winds that play havoc with pedestrians crossing streets near the building. The gustiness and the sudden change of speed and direction of such winds make it difficult for pedestrians to maintain their balance while walking near the building.

Urban street canyon. With many tall buildings lining up the streets of large cities such as Chicago and New York City, the streets often resemble canyons, channeling winds through and causing amplification of winds in constricted areas. The skyscraper-spawned wind and

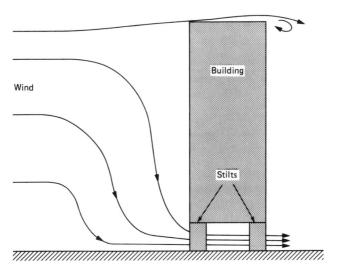

Figure 3.14 Speed-up of wind through stilts.

the urban street canyon generated wind are often coupled together and become indistinguishable from each other. They cause problems at many places in cities on windy days.

Special buildings. A tall building with an arcade piercing through the building such as shown in Figure 3.13 causes a strong wind through the arcade. It can severely restrict building usage and cause damage to the arcade area. An example of such a building that has run into problems is the Earth Sciences Building at the Massachusetts Institute of Technology in Boston. Simultaneous measurements on this building showed that when the wind speed at roof level was 40–50 mph, the wind through the arcade reached 80–90 mph. The cause for this type of wind is the same as for skyscraper-spawned winds, except in this case the high wind directed down the windward face of the tall building is accelerated through the arcade instead of returned to the street; see Figure 3.13(b).

Tall buildings on stilts also have the same problems as do tall buildings with arcades; see Figure 3.14. They generate strong wind beneath the elevated buildings. It should be remembered that all buildings and structures block wind and cause local areas of amplified winds around corners and enhanced turbulence in building wakes. Their effect on neighboring buildings and on pedestrians must be carefully considered, often using wind tunnel tests, in the planning and design of major buildings and structures.

3.4 DIRECTIONAL PREFERENCE OF HIGH WINDS

Extreme wind speeds such as the annual fastest-mile wind reported by most weather stations have preferred wind directions. This is illustrated by the data for Columbia, Missouri, given in Table 2.1. Of the 10 annual fastest-mile wind data listed, 9 were from the northwest, and only 1 was from the southwest. A longer record of the station shows that this preferred northwest direction of high winds for Columbia holds only for data taken prior to 1970. Since then, two preferred high-wind directions appear in the record: northwest and southwest. The change of high-wind direction since 1970 was undoubtedly caused by moving the weather station in 1970 from the old municipal airport, now a city park, to the new regional airport approximately 14 miles south. The new anemometer at the regional airport is surrounded by smooth terrains. In contrast, the old anemometer was at the south edge of the old municipal airport, which had a smooth terrain (the airport) to its north and a rough terrain (the city) to its south. The greater roughness on the south must have reduced the speed of high winds coming from the southwest. This explains why only one preferred high-wind direction—northwest—remains in the old data.

Based on the foregoing example and on limited information published in the literature, for example, Simiu and Scalan (1986), the following tentative conclusions can be drawn on the directional preference of high winds:

1. Extreme winds such as the annual fastest mile or the annual fastest gust at most locations have distinctly preferred directions.
2. The preferred directions are normally in the same directions of the parent storm movements. For instance, in the Midwest of the United States, the summer storms are normally from the southwest and the winter storms are from the northwest. They result in two preferred high-wind directions as reported at the Columbia regional airport. However, other factors such as terrain roughness and topography also affect the preferred extreme-wind directions.
3. Knowing the preferred extreme-wind direction at any location facilitates the development of effective strategies for mitigating wind damage. For instance, if extreme winds are mostly from the northwest, southwest, and west as expected in most places in the Midwest, having a parking lot for cars or aircraft on the east side of

a building can take advantage of the shielding provided by the building. Even better protection will result if the building surrounds not only the west side but also the northwest and southwest sides of the lot.

4. Knowing the preferred extreme-wind direction, a building can be designed with its largest face parallel to that direction. This results in reduced wind loads on the building, and reduced missile damage to claddings. However, other practical considerations may preclude such a design. Besides, this may go against the desire of using the building to protect cars parked outside as mentioned before.

5. Because the high winds at any location may come from different directions, the probability of a certain wind speed being exceeded in a given direction (such as within the southwest octant) is always smaller than the probability of exceeding the same wind speed without considering wind direction. This means that the common practice of disregarding wind direction in the determination of exceedance probability as done in Chapter 2 results in a conservative design.

6. At present (1990), insufficient reliable data on the directional probability of extreme wind speeds exist in the United States for most cities. Studies in Australia indicate that the use of directional wind-speed probability often results in 10–20% reduction in the design wind load. The Australian wind load standard AS-1170.2-1989, one of the best in the world, lists the basic wind speeds for major Australian cities in eight different directions: N, E, S, W, NE, NW, SE, and SW.

REFERENCES

Australian Standard (1989). *SAA Loading Code.* Part 2: *Wind Loads,* Standard Association of Australia, North Sydney.

BOWEN, A. J. (1983). "The Prediction of Mean Wind Speeds Above Simple 2-D Hill Shapes," *Journal of Wind Engineering and Industrial Aerodynamics,* 15, 259–220.

JACKSON, P. S. and HUNT, J. C. R. (1975). "Turbulent Flow over a Low Hill," *Quarterly Journal of the Royal Meteorological Society,* 101, 929–955.

SELVAM, R. P. (1987). "Wind Speed over Irregular Terrain: State of the Art," *Proceedings of the WERC/NSF Symposium on High Winds and Building Codes,*

Kansas City, Missouri, Nov. 2–4, 1987, University of Missouri-Columbia Engineering Extension, 241–248.

SIMIU, E. and SCANLAN, R. H. (1986). *Wind Effects on Structures* (2nd ed.), John Wiley & Sons, New York.

TAYLOR, P. A. and LEE, R. J. (1984). "Simple Guidelines for Estimating Windspeed Variation due to Small Scale Topographical Features," *Climatological Bulletin* (Canada), 18(22), 3–32.

4

Wind Pressure and Forces on Buildings and Other Structures

4.1 INTRODUCTION TO WIND PRESSURE

Definition and Units

The pressure on a surface is the force exerted on the surface per unit area, in a direction perpendicular to the surface. Due to the use of ambient pressure as the reference pressure in wind engineering, the pressure on a surface may be either positive (above ambient) or negative (below ambient). A positive pressure is regarded to act toward the surface, whereas a negative pressure is regarded to act away from the surface. A negative pressure is called a **suction**. Some units of pressure often used are psi (pounds per square inch), dynes per square centimeter, newtons per square meter (pascals), and so on. Due to the relative smallness of wind pressure on structures, the common English unit used for wind pressure is psf (pounds per square foot) rather than psi. The preferred SI unit for wind pressure is pascal (Pa), with the atmospheric pressure measured in bar (10^5 Pa or 14.5 psi) and millibar (10^{-3} bar).

External and Internal Pressures

For hollow structures such as a building or a chimney exposed to wind, there is not only an **external pressure** on the exterior surface of the structure, but an **internal pressure** also exists on the inside surface of the structure. The two may act in the same direction or in opposite directions, depending on their location on the structure and the wind direction. It is the vector sum (algebraic difference) of the two that determines the magnitude and direction of the wind load per unit area at a given location on the structural wall.

Ambient Pressure

The value of wind-induced pressure depends on the reference pressure used. In fluid mechanics, the reference pressure is either the atmospheric pressure or the vacuum. The **gage pressure** is the pressure above the atmospheric pressure, whereas the **absolute pressure** is the pressure above the vacuum. However, in studying wind loads on structures, the reference pressure is the **ambient pressure,** defined as the air pressure at the location of a structure if the structure did not exist to block the wind. The ambient pressure of a prototype building is the atmospheric pressure. In the case of a structure model tested in a wind tunnel, the ambient pressure is the air pressure in the test section, which is often quite different from the atmospheric pressure.

To determine the ambient pressure of a structure in wind accurately, the pressure must be measured at a proper lateral or vertical distance from this structure and from neighboring structures, so that the ambient pressure is not affected by flow separation and wakes generated by structures. Incorrect values of ambient pressure will result if measurements are taken too close to a structure or other obstacles. Taken too far from the test structure will also result in erroneous measurements. For this reason, expert judgment is often needed to determine the best location to measure ambient pressure in the field. For a building surrounded by other structures or obstacles such as trees, it may not even be possible to measure the correct ambient pressure anywhere on the ground. In wind tunnel tests, the ambient pressure is often measured at an appropriate location of the wall of the tunnel test section. It can also be measured by using an ambient tube or the ambient opening of a Pitot tube.

Many prototype measurements (field studies) conducted in the past did not compare well with wind tunnel test results because of

Sec. 4.1   Introduction to Wind Pressure   71

incorrect determination of the ambient pressure outdoor. Some field studies used the internal pressure of a building as the reference pressure, which also created problems. Correct determination of the ambient pressure has been and continues to be one of the most troublesome problems in prototype measurements.

Stagnation Pressure

The only place where the external pressure on a structure can be predicted accurately from theory is at the stagnation point, located slightly above the center of the windward surface. Assuming a steady wind of uniform velocity upstream, application of the Bernoulli equation in fluid mechanics between a point upstream and the stagnation point yields

$$p_s = p_a + \frac{1}{2}\rho V^2 \qquad (4.1)$$

or

$$p_s - p_a = \frac{1}{2}\rho V^2 \qquad (4.2)$$

In the foregoing equations, p_s is the **stagnation pressure** (i.e., the pressure at the stagnation point), p_a is the ambient pressure, ρ is the air density, and V is the velocity (or more precisely the speed) of the upstream wind—the undisturbed or free-stream wind speed.

When the ambient pressure is atmospheric, $p_a = 0$, and

$$p_s = \frac{1}{2}\rho V^2 \qquad (4.3)$$

Equations 4.2 and 4.3 indicate that the rise in pressure from free stream (ambient) to the stagnation point is exactly $\rho V^2/2$. This means we should expect the pressure near the center of the windward surface of a structure to be $\rho V^2/2$ (above the ambient).

Under standard atmospheric conditions (14.7 psia and 60° F), the density of air is 0.00237 slug/ft^3. Therefore, for a 100-mph (147-fps) wind, the pressure near the center of the windward surface is $p_s = \frac{1}{2} \times 0.00237 \times 147^2 = 25.6$ psf. This illustrates how to determine the pressure at a stagnation point. The stagnation pressure, incidentally, is the highest steady-state pressure that can be produced by wind on a structure. In the case of a tall structure exposed to different wind speeds at different heights, the speed V for calculating the stagnation pressure is that at the height of the stagnation point. In the literature, the quantity $\rho V^2/2$ is called not only **stagnation pressure** but also under other names such as **dynamic pressure** and **velocity pressure**.

Dimensionless Pressure (Pressure Coefficient)

Local mean pressure coefficient. The pressure p at an arbitrary point on a structure can be expressed in dimensionless form as follows:

$$C_p = \frac{p}{\frac{1}{2}\rho V^2} \tag{4.4}$$

where C_p is the dimensionless pressure called the **pressure coefficient**. The pressure p is measured above the ambient pressure. The velocity V is that at a reference height, usually the roof height for low-rise buildings and the local height where p is measured for tall structures.

In the study of wind load on structures, pressure distributions around structures are usually presented in dimensionless form, the pressure coefficient. This has an advantage over the use of pressure directly, for dimensionless results are more general in nature. For instance, if pressure instead of dimensionless pressure were plotted in Figures 4.1–4.4, the graphs would be valid only for one particular wind speed, at a particular air density. However, in dimensionless form the graphs are valid for almost any wind speed and air density, as long as the shape of the building and the orientation of the wind are fixed.

From Eqs. 4.3 and 4.4, the pressure coefficient at any stagnation point has a value equal to one. It can be less than 1.0 if we use a reference velocity greater than the free-stream velocity corresponding to stagnation-point height in Eq. 4.4, as when the free-stream velocity at the roof height of a building is used.

The pressure coefficient C_p discussed so far is based on the time-averaged value of p at any given location and the time-averaged value of V. It is to be referred to hereafter either as the **local mean pressure coefficient** or the **local pressure coefficient**, or simply as the **pressure coefficient**. The local mean C_p is needed for determining the variation of pressure over building surfaces.

Sec. 4.1 Introduction to Wind Pressure 73

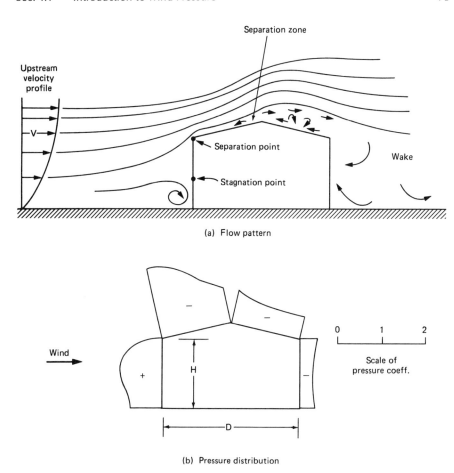

(a) Flow pattern

(b) Pressure distribution

Figure 4.1 Typical wind flow pattern and pressure distribution around a low-rise building having a gable roof of small slope.

Area pressure coefficient. Integrating the local mean pressure coefficient over a surface area such as the roof or the windward wall of a building and then divided by the area yields the **area pressure coefficient**, which can be used conveniently for determining the wind loads on specific areas of buildings. Building codes and standards often list the values of the area pressure coefficient for various parts of buildings. They are often referred to in codes and standards simply as the "mean pressure coefficient" or "pressure coefficient," without distinguishing from the local mean pressure coefficient.

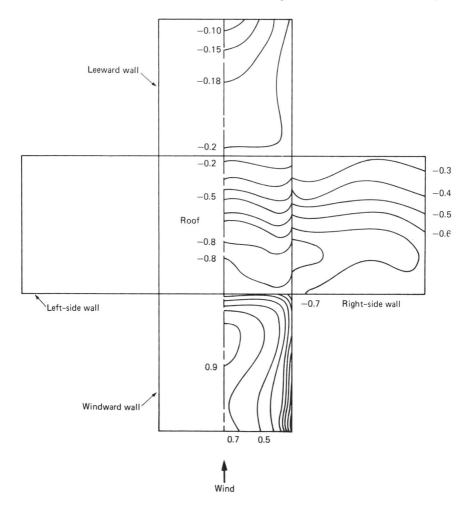

Figure 4.2 Typical distribution of local mean pressure coefficient over a low-rise, flat-roof building in boundary-layer wind.

Peak and RMS pressure coefficients. There are two other pressure coefficients relevant to structural engineers—one based on the peak fluctuating pressure $\hat{p}$ and the other based on the root-mean-square pressure fluctuations p_{rms}. They are defined respectively, as

$$\hat{c}_p \text{ (peak pressure coefficient)} = \frac{\hat{p}}{\frac{1}{2}\rho V^2} \quad (4.5)$$

Sec. 4.1 Introduction to Wind Pressure

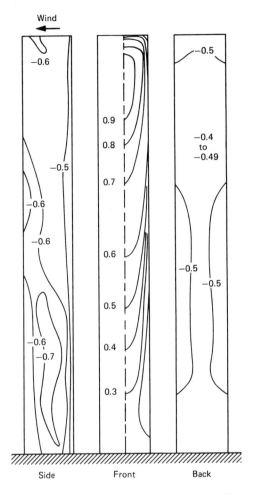

Figure 4.3 Typical distribution of local mean pressure coefficient over a tall building in boundary-layer wind.

and

$$(C_p)_{rms} \text{ (rms pressure coefficient)} = \frac{p_{rms}}{\frac{1}{2}\rho V^2} \qquad (4.6)$$

In both equations, the value of V is still the mean (time-averaged) free-stream velocity. The local values of peak pressure coefficient and the rms pressure coefficient are important for designing structural claddings such as curtain walls.

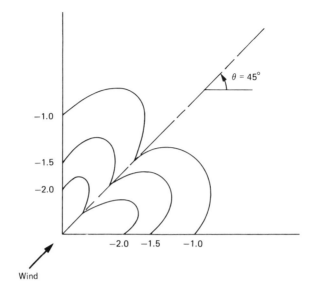

Figure 4.4 Distribution of local mean pressure coefficient in a roof corner facing wind.

4.2 WIND PRESSURE ON RECTANGULAR BUILDINGS

Rectangular buildings are those whose top view (plan) is rectangular or nearly rectangular. The three dimensions of a rectangular building are the **height** H, the **breadth** B, and the **depth** D. The depth is the horizontal dimension parallel to wind, and the breadth is the horizontal dimension normal to wind. In this book, the longer of the two horizontal dimensions, B or D, is referred to as the **length L**, and the shorter of the two as the **width W**. While B and D are determined with wind direction in mind, L and W are determined from building geometry alone without regard to wind direction. For roofs that are not flat, the height H is normally taken as the mean roof height. In general, wind pressure distribution around a rectangular building depends on the ratios D/H, B/H, δ/H, and z_o/H, where δ is the thickness of the atmospheric boundary layer and z_o is the roughness of the ground. Some common characteristics of wind flow patterns around, and pressure distribution over, all rectangular buildings are discussed next.

External Flow and Pressure

Flow pattern around building. Figure 4.1(a) shows the typical wind pattern around a rectangular building. In the neighborhood of the roof and near the side walls (i.e., the two walls parallel to wind), the air movements are highly turbulent, and there is always a backflow near the surface. These regions are called **separation zones**. The pocket of flow trailing the building, called the **wake**, is also highly turbulent. In both the separation zones and the wake, the pressure of air is below the ambient pressure, and strong turbulence is generated to cause fluctuations of the negative pressure (suction). The pressure of air on the windward wall is, on the contrary, higher than ambient. The fluctuations of pressure on the windward wall is mostly due to the turbulence carried in the free stream (approaching flow).

External pressure distribution. Figure 4.1(b) is a graphical representation of the pressure distribution around the same building in (a). The pressure is represented in dimensionless form—the local mean pressure coefficient. Note that the pressure is positive on the windward wall. For this building with a small roof slope, the entire roof, the side walls, and the leeward wall all have negative pressure. For gable roofs with a steep slope, the windward part of the slope experiences positive pressure. Positive pressure simply means above-ambient pressure, whereas negative pressure (suction) means below-ambient pressure.

Generally, the local mean pressure coefficient varies from 1 to 0 on the windward wall, with decreasing values toward the edges of the wall. The average C_p for the windward wall is approximately 0.8. The values of C_p for roofs of small slopes (less than approximately 20 degrees), leeward walls, and side walls are all negative (suctions). The average of C_p for the side walls is approximately -0.7, and the average for the leeward wall is between -0.2 and -0.5, depending on building geometry—mainly the depth-to-breadth and the height-to-breadth ratios. The pressure coefficients for various types of roofs will be discussed later.

A more precise representation of the distribution of local mean C_p over various walls and the roof is shown in Figure 4.2 for a low-rise, flat-roof building. The contour lines in the figure represent lines of constant pressure coefficient. Note that a plus sign represents pressure, whereas a minus represents suction. A similar plot for a high-rise building is shown in Figure 4.3.

The distributions of C_p for buildings and structures of various shapes can be found in publications such as ASCE (1961) and Simiu and Scalan (1986). However, most of the published values of C_p for buildings, such as those given in these two references, were based on wind tunnel measurements conducted many years ago in smooth uniform winds rather than in turbulent boundary-layer winds. They must be viewed with caution and used only if boundary-layer wind tunnel (BLWT) test results are unavailable. Large discrepancies exist between the pressure coefficients measured in boundary-layer winds and those measured in smooth uniform winds. Unfortunately, insufficient BLWT test data exist today (in 1990) for pressure coefficients of buildings of various shapes and dimensions. Therefore, past data collected in wind tunnels of smooth uniform flow may still be needed in situations where better information is unavailable. Some pressure coefficients based on BLWT tests are available in publications such as Jensen and Frank (1963, 1965); Davenport, Surry, and Stathopoulos (1977, 1978); Best and Holmes (1978) for low-rise buildings; and Cermak (1977), Akins and Cermak (1975), and Peterka and Cermak (1974) for high-rise buildings. They have been used for the pressure coefficients specified in contemporary building codes and standards.

The largest suction on buildings exists near roof corners and ridges. When wind is blowing diagonally to the walls of a flat-roof building, the local mean (time-averaged) pressure coefficient in the windward corner of the roof may be of the order of -2. The distribution of suction in a windward corner is as shown in Figure 4.4.

The local values of the peak and rms pressure coefficients are important for designing structural claddings. Some studies have found that the peak pressure coefficient in the roof corner of tall buildings can be almost as high as -10, which produces tremendous local suction on roof corners. This explains the high negative values of pressure coefficients required by modern building codes and standards for designing claddings at roof and wall corners.

Reattachment of streamlines. For a building having a small length (depth) in the wind direction, the entire side wall and the roof of the building are engulfed in separation zones; see Figure 4.1(a) and Figure 4.5(a). In this case, low pressure (high suction) exists throughout the roof and the side walls. In contrast, for an elongated building, the streamlines that separate from the windward edges reattach to the building walls and the roof in a manner as shown in Figure 4.5(b). The phenomenon is called **reattachment**. In this case, separation zones

Sec. 4.2   Wind Pressure on Rectangular Buildings   79

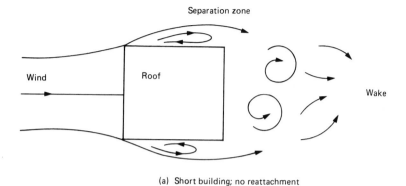

Figure 4.5 Separation and reattachment of flow along buildings of different lengths.

cover only the windward part of the roof and the side walls, from the leading edge of the building to the reattachment point. Upon reattachment, the external pressure recovers from its low values in the separation zones. Much less suction is encountered in the reattachment zone than in the separation zone.

Flow above roof and pressure. The pattern of wind flow above a roof and the resultant external pressure generated on the roof depend in a complex manner on roof geometry, building height, breadth, and depth, and the characteristics of the approaching wind. It is difficult to describe their common behavior. However, a few general rules of fluid mechanics can be applied to such a complex flow field to help understand its behavior. For instance, for both flat roofs and gable roofs with a

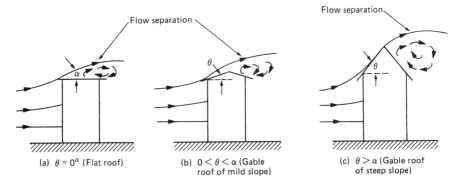

Figure 4.6 Wind flow pattern above flat and gable roofs.

mild slope and with wind perpendicular to the eave and ridge, due to the upward deflection of wind by the windward wall, the entire roof is engulfed in a flow separation zone. The situation is depicted in Figure 4.6(a) and (b). In these two cases, the entire roofs are subject to suction and strong fluctuations of negative pressure. Note that the roof slope in (b) is considered mild because the roof angle θ is smaller than the angle α for the streamline shown in (a). In case (c), the roof angle θ is larger than the angle α in (a). Because streamlines are deflected by the windward half of the roof to bend upward, a pressure rather than suction is generated on the windward half of the roof. In this case flow separation does not occur until streamlines have reached the ridge of the roof. Consequently, only the leeward half of the roof has flow separation.

Note that the streamline separation angle α shown in Figure 4.6(a) is not constant. It depends on building geometry—height, breadth, and depth. The higher a building is, the steeper the streamlines are bent by the windward wall, and the larger α becomes. This explains why the roof pressure coefficient C_p given in Table 4.1 becomes positive at increasing roof angles θ as the building height increases. For instance, at H/D less than or equal to 0.3, the pressure coefficient becomes positive at a roof angle of approximately 20 degrees, whereas when H/D is increased to 1.5 the pressure coefficient does not become positive until θ exceeds 40 degrees.

When wind velocity is parallel to the ridge line, regardless of the roof angle, the flow separates from the windward edge in a manner similar to case (a) in Figure 4.6. Consequently, the pressure coefficients are always negative, as indicated in Table 4.1. Note that the values of C_p given in Table 4.1 are area-averaged values. They do not mean that

TABLE 4.1. Pressure Coefficient C_p for Gable Roof Used in ANSI A58.1-82 and ANSI/ASCE-7-1988

Wind Direction	Roof Height-to-Depth Ratio H/D	Pressure Coefficient, C_p				Leeward Part of Roof
		Windward Part of Roof Roof Angle θ (degrees)				
		0	20	40	≥ 60	
Normal to ridge	≤ 0.3	-0.7	0.20	0.40	0.01θ	-0.7 for all cases
	0.5	-0.7	-0.75	0.30	0.01θ	
	1.0	-0.7	-0.75	0.30	0.01θ	
	≥ 1.5	-0.7	-0.90	-0.35	0.01θ	
Parallel to ridge	H/B or $H/D \leq 2.5$	-0.7				-0.7
	H/B or $H/D > 2.5$	-0.8				-0.8

Notes:
1. Each value of C_p represents the average value over an area such as the windward part of the roof.
2. H is the average height of the roof above the ground level, B is the breadth of the roof being horizontal and perpendicular to the wind, and D is the depth of the roof in the direction of the wind.
3. The wind speed in calculating C_p is that at mean roof height H.

pressure is constant over the entire roof or even over half of the roof. Spatial variation of C_p still exists within such areas.

The behavior of pressure distribution over other types of roofs such as round roofs and hip roofs can be explained in a similar manner. Generally, hip roofs create smaller suction in corners than gable roofs. Consequently, hip roofs are less vulnerable to high winds than are gable roofs. The strongest suction generated on flat roofs and gable roofs exists when wind is diagonal to buildings; see Figure 4.4.

The foregoing discussion pertains to the roofs of buildings. For a **freestanding roof,** that is, a roof on columns or other supports without enclosed walls under the roof, wind passes over both sides (above and below the roof), and there is no windward wall to deflect the wind upward or sideways. Therefore, freestanding roofs (or **free roofs** in brief) behave quite differently from the roofs of buildings. For instance, for a freestanding flat (horizontal) roof, the wind flows above and beneath the roof are approximately symmetric. They create not much uplift. However, due to the development of a turbulent boundary layer on both sides of the roof as shown in Figure 4.7(a), large pressure fluctuations still exist on both sides of the roof, which may cause vibration of the roof if the roof is flexible or not tied down well.

In case (b) of Figure 4.7, the roof is leaning in the leeward direction ($0° < \theta < 90°$). Flow separates from both the top and bottom edges of the roof, causing a turbulent wake on the downwind side of the roof. In this case, the pressure above the roof is higher than that underneath it, and hence the roof experiences a net thrust (lift) downward. In case (c), the roof is leaning in the windward direction ($90° < \theta < 180°$). The flow pattern is opposite to that of a leeward-leaning roof, and the net thrust (lift) on the roof is upward. Case (d) is a free gable roof. For the windward half of the roof, the pressure is higher on the upper side than the underside, causing a net downward thrust on the windward roof. The opposite happens on the leeward half of the roof. The downward thrust on the windward roof and the upward thrust on the leeward roof

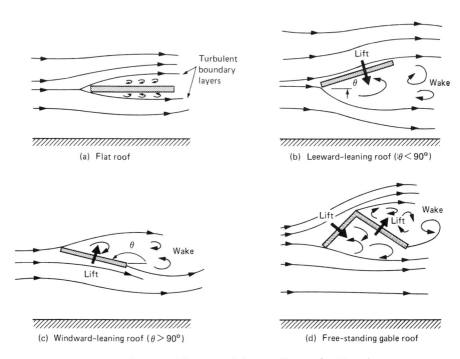

Figure 4.7 Wind flow around free-standing roofs. (Note that supports for roofs are not shown.)

causes a counterclockwise torque on the entire roof for wind blowing from the left as shown in the sketch. In all four cases, the roofs are subjected to the action of turbulence and large fluctuating loads.

Effect of architectural features. Architectural features such as canopies, roof overhangs, awnings, carports, parapets, and fences, all create unique wind effects. The disturbance to wind generated by these features can be analyzed or understood by applying the same fluid mechanics principles as described before. For instance, for any roof overhang facing a wind (i.e., on the windward side), a stagnation pressure will be generated beneath the overhang. This causes a large upward lift force on the underside of the overhang. If the roof is horizontal or has a small slope—case (a) and (b), Figure 4.6—high suction will be generated above the overhang. This suction lift from above, combined with the pressure lift acting on the underside of the overhang, causes a large net uplift on the overhang which may damage the overhang in high winds.

A canopy near the roof level of a house behaves like a roof overhang. On the other hand, a canopy in the middle or near the bottom of a tall building, when existing on the windward side, will encounter a strong downward draft caused by the building—see the discussion of skyscraper-spawned wind in Section 3.3. The downdraft causes stagnation pressure on the upper side of the canopy whereas flow separation beneath the canopy causes suction on the underside. Consequently, a net downward thrust is generated on the canopy.

The value of parapets in reducing wind effect on roofs appears overrated. According to a recent study (Stathopoulos and Baskaran 1985), even though parapets somewhat reduce the high suction on roof edges and slightly increase the lower suctions on the interior areas of roofs, they may significantly increase the suctions in roof corners if the parapets are low, say, less than 1 m. High parapets, though more effective in reducing wind load on roofs, may be expensive and impractical.

Fences and windbreaks are effective in reducing windload on buildings if they are sufficiently high and dense, and placed sufficiently close to a building. When placed at a distance from a building, their wakes may cause fluctuating pressure on the building higher than without the fences or windbreaks.

Effect of nearby structures. The bombardment of a structure by the turbulence created by a nearby upwind structure is called **buffeting**. Tall flexible buildings are strongly susceptible to wind-induced vibration caused by buffeting from neighboring tall buildings. Tall buildings also bring high winds from high elevation to low heights, causing increased wind load on low buildings nearby. In contrast, low buildings surrounded by other low buildings have a minimum impact on each other. In fact, they may even reduce the maximum wind loads on each other.

Pressure Inside Buildings: The Internal Pressure

Internal pressure coefficient. Wind forces on various parts and components of buildings depend on both the external and the internal pressures. It is the vector sum of the two that determines the magnitude and the direction of the resultant force on each part of building cladding. It is important to know how the internal pressure behaves. As with external pressure, a dimensionless internal pressure can be defined as follows,

$$C_{pi} = \frac{p_i}{\frac{1}{2}\rho V^2} \quad (4.7)$$

where C_{pi} is the dimensionless internal pressure, also called the **internal pressure coefficient,** and p_i is the internal pressure above ambient.

Change of internal pressure with openings. The internal pressure behaves differently from the external pressure in that it is usually uniform throughout a building, except for tightly isolated rooms that cause a different pressure in each room. Another difference is while external pressure is not much affected by window or door openings, opening or closing a window or door has a drastic effect on the internal pressure.

The variation of the internal pressure of a building with opening conditions is illustrated in Figure 4.8. For simplicity, the wind is assumed to be steady, that is, not varying with time. In (a) of the figure, a completely hermetic building is assumed. In this case, the internal pressure is independent of the wind. It can have any value (either positive or negative) depending on the initial conditions. In (b), a single opening exists on the windward side, and no openings or leakages exist elsewhere. Regardless of the size of this windward opening, the steady-state pressure inside becomes as high as the external pressure on the windward wall near the opening. In this case, the forces due to internal

Sec. 4.2   Wind Pressure on Rectangular Buildings   85

pressure are in the same directions as those due to external pressure for the roof and the three suction walls. This causes large wind loads on the roof and the suction walls.

In (c), the only opening is on a suction cladding—leeward wall, side walls, or roofs. The steady-state pressure inside is suction, having the same magnitude as the external pressure at the suction opening. This causes a large wind load on the windward wall, and reduced loads on the suction sides. The **suction sides** are defined as the four sides having negative external pressure, namely, the leeward wall, two side walls, and the roof. Finally, (d) represents the case where openings exist on both windward and the suction sides. The internal pressure can be either positive or negative, depending on whether greater openings exist on the windward wall or the suction sides. When greater openings exist on the windward wall, the internal pressure will be positive. On the other

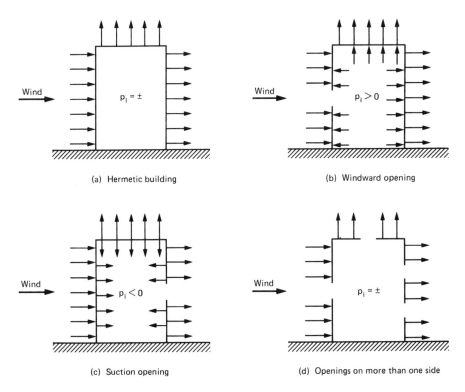

Figure 4.8 Variation of internal pressure with exterior openings for a typical building. (From Liu and Saathoff, 1982.)

hand, when greater openings exist on suction sides, the internal pressure will be negative. Note that (d) causes strong natural ventilation (wind blowing through the building), especially when the windward and suction openings are both large. In all the cases with a single opening, the steady-state internal pressure is the same as the steady external pressure at the opening. Therefore, if the only opening exists at the stagnation point, the steady internal pressure will be as high as the stagnation pressure. If the only opening is near a roof corner, the same high suction at the roof opening will prevail inside the building.

The variation of internal pressure with building openings as described in the previous paragraph has profound implications. For instance, if during high wind, a person opens a windward door or window of a house having a lightweight roof, he or she greatly increases the chance that the roof will be lifted off due to increased internal pressure. The loss of the roof may in turn cause wall failure and the destruction of the entire house. Likewise, once a windward door or window is forced open by high winds, or broken by windborne debris, the house may be endangered in the same manner. Field investigators following hurricane and tornado disasters have found that this appears to be a common mode of failure for wood-frame houses and buildings that have unreinforced masonry walls (concrete-block walls). This points to the importance of designing buildings, especially those with light roofs and unreinforced masonry walls, against the high internal pressure that may be generated by windward window or door failure, unless the designer is certain that the rise of internal pressure will cause a window or door on a suction side to fail before the roof or walls fail. Although a building may be fitted with specially designed blow-out panels to relieve the high internal pressure generated by high winds, one should be cautioned that the activation of such devices causes a strong wind penetrating through the building which may cause costly damage to the building interior.

Predicting steady internal pressure. The equilibrium or steady-state internal pressure of a building can be predicted once the wind speed, the wind direction, and the sizes and locations of the building openings are known.

First, consider the case of a building with an **open interior.** This means the building either has only a single room or has more than one room with large openings between rooms. In either case, the pressure inside the building is everywhere the same; only a single internal pressure exists for such a building. Suppose the building has N exterior openings distributed arbitrarily over the building. By treating each

Sec. 4.2 Wind Pressure on Rectangular Buildings

opening as an orifice through which air flows, the volumetric discharge of wind entering or leaving the jth opening is

$$Q_j = C_d A_j V_j = \pm\, C_d A_j \sqrt{2\,|p_j - p_i|/\rho} \tag{4.8}$$

where C_d is the discharge coefficient of the orifice which has a value somewhat less than one (approximately 0.73 for a sharp-edged square orifice), A_j is the area of the jth opening, V_j is the mean speed of the air jet passing through the orifice, p_j is the external pressure at the orifice location, p_i is the internal pressure, and ρ is the air density. Note that the plus sign is used in Eq. 4.8 when air is entering the building, and the minus sign is used when air is leaving the building. The absolute sign in Eq. 4.8 prevents imaginary numbers produced by negative numbers under the square-root sign. Both Q_j and V_j are considered positive when air enters the building and negative when air leaves the building.

From the continuity equation,

$$\sum_{j=1}^{N} Q_j = 0 \tag{4.9}$$

which means that under steady-state conditions, the rate of air flow entering a building must equal the rate of air flow leaving the building at any time.

A special case of an open-interior building is one with only two openings. If air enters the building through opening 1 and leaves the building through opening 2, Eqs. 4.8 and 4.9 yield (Liu 1975),

$$C_{pi} = \frac{C_{p2} + a^2 C_{p1}}{1 + a^2} \tag{4.10}$$

where C_{pi} is the internal pressure coefficient, C_{p1} and C_{p2} are the external pressure coefficients at openings 1 and 2, respectively, and a is the area ratio A_1/A_2 of the two openings.

Although Eq. 4.10 was derived for a building with only two openings, it can be used as an approximation for an open-interior building with any number of openings, provided that A_1 represents the total area of windward openings, A_2 represents the total area of suction openings, and C_{p1} and C_{p2} represent the average values of external pressure coefficients on windward and suction sides, respectively. Ordinarily, the average values of C_{p1} and C_{p2} are, respectively, 0.8 and -0.5, approximately. For a building with uniform openings over its four walls and the roof, a is equal to one-fourth. Substituting these values into Eq. 4.10

yields $C_{pi} \approx 0.44$. This shows that for a building with uniform openings, the internal pressure is expected to be strongly negative.

For the case of a single opening on the windward wall, Eq. 4.10 reduces to $C_{pi} = C_{p1}$, or the internal pressure becomes the same as the external pressure at the windward opening. When a single opening exists on a suction side, Eq. 4.10 reduces to $C_{pi} = C_{p2}$, or the internal pressure becomes the same as the external pressure at the suction opening. It follows that large internal pressure or suction can be generated by having a single opening or a dominant opening on a building. For instance, if a dominant opening exists on the windward wall near the stagnation point, the mean internal pressure coefficient can approach 1.0. Such high internal pressure can have serious consequences to buildings that have lightweight roofs and/or unreinforced masonry walls.

The foregoing analysis for predicting internal pressure holds not only for large openings such as doors and windows, it also holds for small openings such as the cracks around windows and doors, as long as the cracks are large enough to generate turbulent flow through them. However, if the cracks are so small that the flow through them is laminar, Eq. 4.8 must be changed to

$$Q_j = \frac{b_j}{12\mu}(p_j - p_i) \quad (4.11)$$

where

$$b_j = \frac{L_j W_j^3}{d_j} \quad (4.11a)$$

Note that μ is the dynamic viscosity of air and L_j, w_j, and d_j are, respectively, the length, width, and depth of the jth crack.

The equivalent of Eq. 4.10 for two cracks is

$$C_{pi} = \frac{C_{p2} + bC_{p1}}{1 + b} \quad (4.12)$$

where $b = b_1/b_2$ can be calculated from Eq. 4.11a.

Eq. 4.12 for two cracks can again be used as an approximation to determine the internal pressure of buildings having any number of cracks, provided that b_2 and C_{p2} represent average values for the suction sides (i.e., leeward and side walls, and the roof). For instance, if we assume that cracks are uniformly distributed over all sides (four walls plus the roof) of a building and that the crack width w and crack depth d are constant for all cracks, we have $b = b_1/b_2 = L_1/L_2 = 1/4$. Using

Sec. 4.2 Wind Pressure on Rectangular Buildings

the values $C_{p1} = 0.8$ for windward and $C_{p2} = -0.5$ for suction sides, Eq. 4.12 yields $C_{pi} = -0.24$.

The foregoing calculations show that depending on how large cracks and openings are or whether the flow is laminar or turbulent, for a building with uniform distribution of openings and/or cracks, the internal pressure coefficient generated by wind is expected to be between -0.2 and -0.4. This means that high winds can generate a large negative pressure inside any building of uniform openings or porosity. Note that meteorologists have for years been measuring atmospheric pressure indoors. Such practice can result in large errors in the atmospheric pressure measured in high winds. Liu and Darkow (1989) discussed and treated this problem with attempts being made to correct the problem. Building codes and standards also have traditionally assumed that the internal pressure was nearly zero with uniform distribution of openings, which is an unsafe assumption. Note that it was not until 1972 that ANSI A58.1 and model codes in the United States began even to consider wind loads generated by internal pressures.

Blast effect of internal pressure. As soon as a windward window or door is broken in high winds, outside air rushes into the building, causing the internal pressure to rise rapidly. Due to the inertia of air, the internal pressure will not only reach the external pressure at the opening but will overshoot for a short while—a blast effect. From theoretical analyses (Liu and Saathoff 1981), the sudden opening of a large windward window can easily cause the internal pressure to rise momentarily to a value 50% higher than the external pressure at the opening. This means that the breakage of a large windward window near the stagnation point may cause the internal pressure coefficient to exceed 1.5. This blast effect, yet to be demonstrated by experiments, is believed to play an important role in causing wind damage to small buildings that have weak roofs and/or weak walls, such as single- or double-family houses.

Helmholtz oscillation of internal presssure. Both wind tunnel experiments and theoretical studies have revealed that when a building exposed to wind has a single opening, regardless of the wind direction an oscillation of internal pressure may occur. The phenomenon is the same as the **Helmholtz resonance** discussed in acoustics. The frequency of this oscillation can be calculated from

$$n_H \text{ (Helmholtz frequency)} = \frac{A^{1/4}}{\pi^{5/4}} \sqrt{\frac{k\, p_a}{2\rho \forall}} \qquad (4.13)$$

where A is the area of the opening, $\mathbb{V}$ is the internal volume (air volume) of the building, P_a is the atmospheric pressure (absolute pressure), ρ is the air density, k is the adiabatic exponent of air equal to 1.4, and $\pi = 3.1416$.

Example 4-1. Consider a building having an internal volume of 500 m³ and a window opening of 1 m². The atmospheric pressure is 1.013×10^5 Pa (absolute), and the air density is 1.1 kg/m³. Find the Helmholtz frequency of this building.

[Solution]. For this building $A = 1$ m², = 500 m³, $k = 1.4$, $p_a = 1.013 \times 10^5$ Pa, and $\rho = 1.1$ kg/m³. Substituting these values into Eq. 4.13 yields $n_H = 2.7$ Hz. This shows that the Helmholtz frequency of this building with a 1-m² opening is 2.7 hertz.

Although large Helmholtz oscillations of internal pressure have been observed in wind tunnel studies, fortunately they have not been found to be serious in the limited number of field investigations of prototype buildings conducted to date to determine internal pressure fluctuations. It appears that due to relatively large air leakage, cladding flexibility, and structural damping, little Helmholtz oscillation exists in prototype buildings under ordinary conditions, even when a large door or window is open. Therefore, the Helmholtz oscillation need not be considered in ordinary structural design. However, Helmholtz oscillation may cause problems to buildings in rare cases such as

1. When a large single opening exists on a building and when the external pressure at this opening fluctuates at a dominant frequency near the Helmholtz frequency. Such an oscillating external pressure can be generated by vortex shedding from an upstream structure.
2. When a large single opening exists on a building and when the Helmholtz frequency matches the natural frequency of certain cladding components of the building. This can happen to buildings having flexible claddings such as a sheet metal roof.

In addition, those who conduct wind tunnel tests of internal pressure fluctuations must make sure that Helmholtz oscillation of internal pressure in models will not misrepresent the behavior of prototype buildings.

Correlation of Pressure Fluctuations

A good correlation often exists for pressure fluctuations over different areas of a building. Such correlation can cause large loads and sometimes serious vibration of the building and its claddings. For instance, as the pressure at a given point of a wall rises (or falls), the pressure at another location of the same wall also rises (or falls) in synchrony or in phase. This produces a large peak load on the entire wall and can cause the wall to vibrate if the frequency of the load fluctuations is in the neighborhood of the natural frequency of the wall. A second example is the good correlation between the pressure on the windward wall and the suction (negative pressure) on the leeward wall. This produces a large drag on the building and may cause the entire building to vibrate in the along-wind direction. A third example is the good correlation between internal pressure and external suction. As the internal pressure of a building having a large windward opening rises (or falls) in wind, the external suction on the suction sides (i.e., leeward and side walls, and the roof) also rises (or falls) in phase. This again creates large loads on claddings and possible cladding failure or vibration. The good correlation between windward external pressure and the internal pressure produced by a windward opening, however, reduces rather than increases the load on the windward wall.

To understand why the pressure fluctuations at various parts of a building are well correlated, one must realize that the pressure fluctuations on a building are caused by two sources of turbulence: the **free-stream turbulence** carried in the approaching wind and the **signature turbulence** generated by the building itself. While the signature turbulence has a length scale smaller than the building, the free-stream turbulence often has a length scale much larger than the longest dimension of the building. This large-scale, slowly varying, free-stream turbulence, when buffeting on the building, produces good correlations in pressure fluctuations.

When a building encounters a slowly varying large eddy in the wind, the flow around the building at any given time t can be regarded as **quasi-steady**. In a quasi-steady flow, the pressure p at any location varies with the instantaneous velocity V in the same manner as for steady flow, namely,

$$p(t) = C_p \frac{\rho V^2(t)}{2} \qquad (4.14)$$

where $p(t)$ is pressure at time t and $V(t)$ is the free-stream velocity at t. Because C_p does not vary with time in a quasi-steady flow, the pressure $p(t)$ for any location on a building in a quasi-steady flow varies with the square of the instantaneous free-stream velocity $V(t)$.

For the pressure at two different points 1 and 2 on a building, Eq. 4.14 yields

$$p_1(t) = C_{p1} \frac{\rho V^2(t)}{2} \quad \text{and} \quad P_2(t) = C_{p2} \frac{\rho V^2(t)}{2} \quad (4.15)$$

where subscripts 1 and 2 denote points 1 and 2, respectively. From Eq. 4.15,

$$\frac{p_1(t)}{p_2(t)} = \frac{C_{p1}}{C_{p2}} = K \text{ (constant)}$$

or

$$p_1(t) = K p_2(t) \quad (4.16)$$

Equation 4.16 shows that in an idealized quasi-steady flow without signature turbulence, the pressure fluctuations at any two points on a building must be linearly proportional to each other. This explains why large-scale free-stream turbulence causes good spatial correlations in pressure fluctuations.

4.3 WIND PRESSURE ON OTHER STRUCTURES

Circular Cylinders

The flow pattern of incompressible flow (liquid or gas) around a long circular cylinder perpendicular to the flow depends on the Reynolds number of the flow defined as

$$\text{Re} = \frac{\rho V D}{\mu} \quad (4.17)$$

where Re is the Reynolds number, ρ is the density of the fluid, V is the velocity of the fluid relative to the cylinder, D is the cylinder diameter, and μ is the dynamic viscosity of the fluid.

The flow pattern around a cylinder in different ranges of Reynolds number important to structural engineering is shown in Figure 4.9. For a

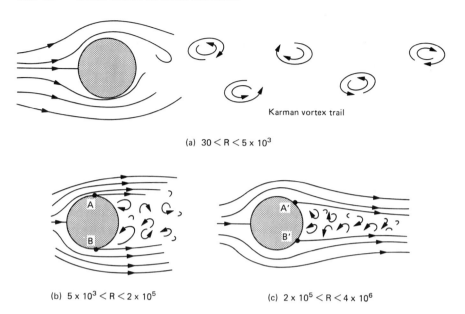

Figure 4.9 Change of flow pattern around circular cylinder with Reynolds number.

Reynolds number greater than approximately 30, but less than about 5000, regular shedding of vortices from the two sides of the cylinder occurs, forming what is called a **von Karman vortex street** or **vortex trail** downstream from the cylinder as shown in Figure 4.9(a). The flow in this case is unsteady but laminar. As the Reynolds number exceeds 5000 approximately, the wake downstream of the cylinder becomes turbulent, whereas the flow around the cylinder upstream of the wake remains laminar. The wake becomes more and more turbulent as the Reynolds number increases. Before reaching the **critical Reynolds number** (2×10^5), flow separation occurs at the two sides of the cylinder—points A and B in (b). The width of the wake is rather wide—wider than the cylinder diameter—and vortex shedding is rather regular. As the Reynolds number exceeds 2×10^5, the flow separation points suddenly shift downstream—from A and B to A' and B', respectively, as shown in (c). This causes a narrower wake, and a sudden decrease in drag as will be discussed later. The vortex shedding during this stage becomes rather random. Finally, when the Reynolds number exceeds approximately 4×10^6, the vortex shedding restores some regularity.

From the foregoing discussion, it can be seen that there are three distinctly different ranges of Reynolds number with distinctly different characteristics in vortex shedding. They are

Range	Re	Shedding Characteristics
Subcritical	30 to 2×10^5	Regular (constant frequency)
Supercritical	2×10^5 to 4×10^6	Random (variable frequency)
Hypercritical	$> 4 \times 10^6$	Regular (constant frequency)

Note that in the literature of vortex shedding, the supercritical range is sometimes referred to as the **critical range,** and what we call hypercritical here is sometimes referred to as **transcritical.** Beware of this difference in notation. The terms used in this book are less mislead-

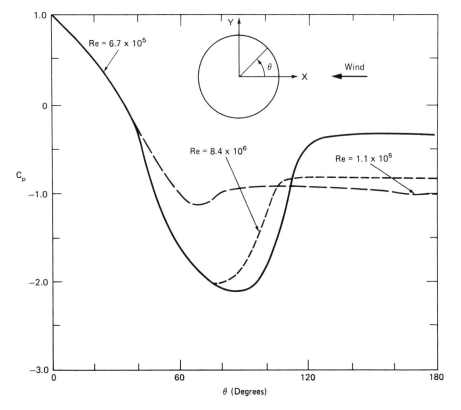

Figure 4.10 Distribution of local mean pressure coefficient around a circular cylinder at different Reynolds numbers. (From Roshko, 1961).

ing since they are consistent with the use of the terms **subsonic, supersonic** and **hypersonic** in studying compressible flow.

Due to the variation of flow patterns around a cylinder with Reynolds number, the distribution of the local mean pressure coefficient is also expected to differ with different Reynolds numbers. This is illustrated in Figure 4.10 for three different Reynolds numbers in the three ranges listed above.

From Figure 4.10, the pressure coefficient at the windward stagnation point is 1.0 at all Reynolds numbers, and it decreases toward a minimum pressure (maximum suction) at approximately 60 degrees for subcritical range, 90 degrees for supercritical range, and 80 degrees for hypercritical range. After reaching its minimum, C_p rises again until it levels off at approximately $\theta = 120$ degrees in supercritical range (at smaller θ in subcritical and hypercritical ranges). A minimum pressure coefficient (maximum suction) approximately equal to -2.0 can be generated at $\theta = 80$ to 90 degrees for flow in the supercritical and hypercritical ranges.

The foregoing results are based on a smooth cylinder exposed to a uniform flow (constant velocity of approach) without free-stream turbu-

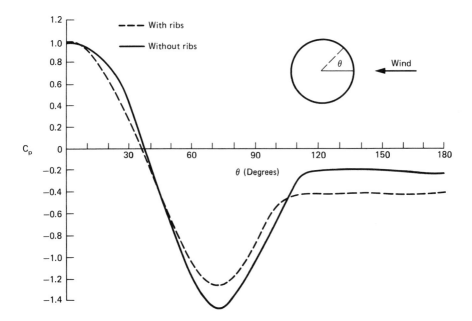

Figure 4.11 Distribution of local mean pressure coefficient around the throat of a hyperbolic cooling tower. (From ASCE, 1987.)

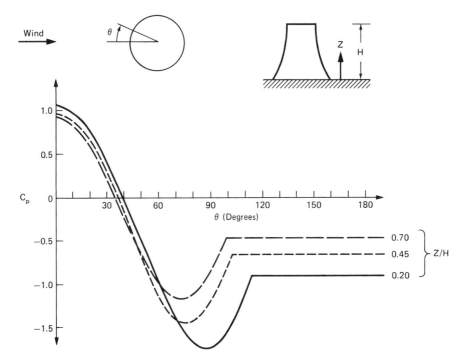

Figure 4.12 Distribution of local mean pressure coefficient at different heights around a hyperbolic cooling tower. The power-law velocity coefficient for this case is $\alpha = 0.27$. (From Proepper and Welsch, 1979.)

lence. If turbulence is present in the free-stream or if the cylinder surface is rough, the critical Reynolds number will be less than 2×10^5 and the transition from subcritical to supercritical and then to hypercritical will not be as distinct as described here. The variation of C_p around a cylinder depends on not only the Reynolds number but also the surface roughness and the turbulence characteristics of the free stream, such as the relative intensity of turbulence and the ratio between turbulence length scales and the cylinder diameter. Furthermore, if the cylinder is placed across a boundary layer flow as in the case of a chimney exposed to wind, the variation of the velocity along the cylinder reduces the coherence of vortex shedding. This again changes the flow pattern and the pressure coefficient around the cylinder.

Hyperbolic Cooling Tower

The pressure distribution around a hyperbolic cooling tower is similar to that around a circular cylinder. In the study of wind load on

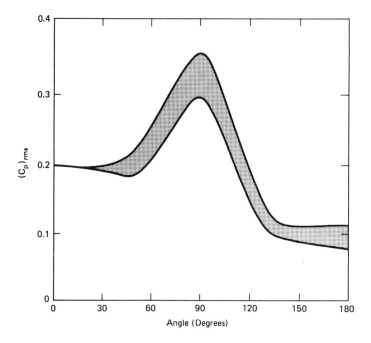

Figure 4.13 Distribution of root-mean-square pressure coefficient around the throat of a hyperbolic cooling tower. (From ASCE 1987.)

cooling towers, it is common to use C_p to represent the difference between the external and the internal pressure coefficients. Figure 4.11 gives the variation of C_p around the throat of typical hyperbolic cooling towers. While the solid line represents towers with smooth surface, the dashed line represents towers with vertical ribs. As will be discussed later, hyperbolic cooling towers are often fitted with vertical ribs to reduce dynamic loads on the tower wall. The curves shown in Figure 4.11 represent average values derived from several field studies (full-scale measurements) described in ASCE (1987). The variation of C_p with height z is given in Figure 4.12.

Since large pressure fluctuations exist on cooling towers, one must consider fluctuating pressure in addition to mean pressure in the design of cooling towers. The intensity of pressure fluctuations on a cooling tower can be characterized by the rms pressure coefficient $(C_p)_{rms}$ defined in Eq. 4.6. Based on limited full-scale test results, the value of $(C_p)_{rms}$ around the throat of hyperbolic cooling towers can be determined from Figure 4.13. More about wind load on hyperbolic towers can be found in ASCE (1987).

4.4 WIND FORCES AND MOMENTS

Cladding Forces

Suppose the average external and internal pressures over the surface of a cladding are p_e and p_i, respectively. The total force on the cladding is

$$F = (p_e - p_i) A \qquad (4.18)$$

where A is the cladding area.

In terms of average external and internal pressure coefficients, C_{pe} and C_{pi} respectively, Eq. 4.18 can be written as

$$F = A (C_{pe} - C_{pi}) \frac{\rho V^2}{2} \qquad (4.19)$$

In using Eqs. 4.18 and 4.19, attention should be given to the signs of p_e, p_i, C_{pe}, and C_{pi}. They are all positive for pressure and negative for suction. The sign of F also bears special meaning: it is positive for a force pushing the wall or the roof inward and negative for an outward force on the building. For example, if for the windward wall of a building $C_{pe} = +0.8$ and $C_{pi} = -0.5$, then $C_{pe} - C_{pi} = 0.8 - (-0.5) = 0.8 + 0.5 = 1.3$. When $\rho = 0.0022$ slug/ft^3, $V = 146.7$ fps, and $A = 10,000$ ft^2, Eq. 4.19 yields $F = 1.3 \times 0.0022 \times 146.7^2 \times 10,000/2 = 307,000$ lb. This yields an inward force of 0.307 million pounds on the windward wall.

Drag Force

Drag is the total thrust on an object produced by flow, in the direction of the flow. The drag force on an object immersed in flow, such as a building or another structure in wind, can be determined from

$$F_D = C_D A_o \frac{\rho V^2}{2} \qquad (4.20)$$

where A_o is a characteristic area of the object such as the area projected onto a plane normal to the flow and C_D is the **drag coefficient**—a dimensionless number that depends on the geometry of the object and the Reynolds number. Eq. 4.20 can be used to determine the drag coefficient of any structure, once we know F_D, A_o, ρ, and V.

The wind-induced drag force on a building can be determined from

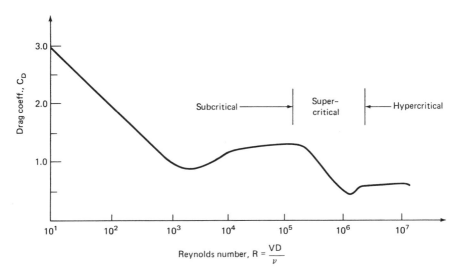

Figure 4.14 Variation of drag coefficient with Reynolds number for circular cylinder in uniform, smooth flow.

Eq. 4.20, with A_o being the projected area of the building on a plane normal to the wind. The drag coefficient of block-type buildings, with wind flowing in a direction normal to one wall, is usually in the neighborhood of 1.3.

Alternatively, the drag force on a building can be calculated from

$$F_D = F_1 + F_2 + F_R \quad (4.21)$$

where F_1 and F_2 are the forces on the windward and the leeward walls and F_R is the drag on the roof. F_R is negligible for flat roofs or roofs of small slopes unless the depth (i.e., the length of the building in the wind direction) is unusually large.

The drag coefficient of a circular cylinder perpendicular to a smooth wind is shown in Figure 4.14. The sudden drop of C_p at Re = 2×10^5 corresponds to the sudden shift of the separation points on the cylinder when this critical Reynolds number is reached. The drag coefficient of a rectangular cylinder at high Reynolds number is a function of the depth-to-breadth ratio of the cylinder D/B. As shown in Figure 4.15, the drag coefficient first increases as D/B increases from zero, reaching a maximum of about $C_D \approx 3.0$ when D/B is approximately equal to 0.5. Further increase of the depth of the cross section causes a decrease of C_D.

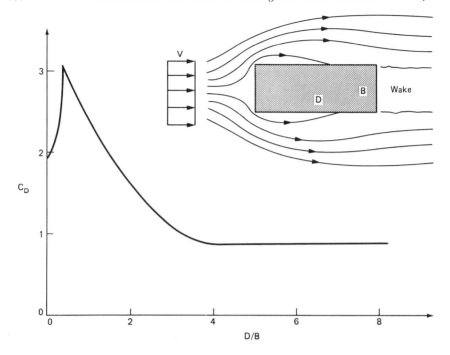

Figure 4.15 Variation of drag coefficient of rectangular cylinder with aspect ratio of cross section, D/B. (From ASCE, 1987.)

Example 4-2. A telephone pole has a diameter of 10 in. and a height of 30 ft. What is the steady-state drag force on the pole when the wind speed is 80 mph?

[Solution]. Assuming standard atmospheric condition, the dynamic viscosity of the air is approximately $\mu = 4.0 \times 10^{-7}$ lb-sec/ft^2. The density of the air is $\rho = 0.00237$ slugs/ft^3. The wind speed is 80 mph = 117.4 fps, and the cylinder diameter is 10 in. = 0.833 ft. Therefore, the Reynolds number of the flow around the cylinder is Re = $\rho V D/\mu$ = $0.00237 \times 117.4 \times 0.833/(4 \times 10^{-7})$ = 5.8×10^5. From Figure 4.14, the Reynolds number is in the supercritical range with a drag coefficient $C_D = 0.8$, approximately. Finally, from Eq. 4.20, $F_D = 0.8 \times (30 \times 0.833) \times 0.00237 \times 117.4^2/2 = 327$ lb.

For open structures with very large porosity or large spacings between neighboring members, as in the case of a wire fence normal to wind, the drag on the entire structure can be determined accurately by adding the drag of each member. This approach produces greater and greater errors when the members of the structure are closer and closer to each other, because the flow around members will affect each other when the clearances between members are small. In such a case, as for a trussed tower, the total drag can be determined directly from Eq. 4.20 applied to the entire structure, with A_o being the gross area of the structure projected on a plane normal to wind. The drag coefficient C_D used in this case will depend on the **solidity ratio** which is the ratio of the projected solid area to the projected gross area. It will also depend on the shape and the roughness of the members, and on the Reynolds number of the members, especially if the members have round or curved surfaces.

The drag coefficients for various other two- and three-dimensional objects can be found in various publications such as ASCE (1961). Note that the dependence of drag coefficient with Reynolds number is much less for cylinders having plane surfaces (e.g., rectangular cylinders) than cylinders having curved or round surfaces. For the former, the drag coefficient is essentially constant over a wide range of Reynolds numbers.

Lift Force

The lift force, simply called **lift**, is the force developed on an object in a direction perpendicular to flow. From this definition, lift and drag are mutually perpendicular. The term **lift** does not mean, as its lay meaning may suggest, an upward force. It can be upward, downward, sideways or in any other direction perpendicular to wind.

Steady-state lift. It should be realized that while drag exists on all objects, there is no steady-state or mean lift on any symmetric object with a symmetric flow around it, such as a sphere immersed in a uniform flow, or a circular cylinder perpendicular to a uniform flow. The symmetric distribution of steady-state forces on the object in the across-flow direction cannot cause steady net lift on such objects.

The only object that has no steady lift regardless of flow direction is a sphere. A circular cylinder encounters no steady lift as long as the cylinder is perpendicular to the flow. Other objects develop lift if the wind is not parallel to their axes of symmetries or if the objects are asymmetric. For instance, consider the rectangular building in Figure 4.16. Due to an oblique wind having an **angle-of-attack** α, both a drag F_D and a horizontal lift F_L are developed as shown in the figure.

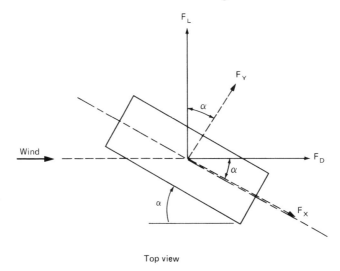

Figure 4.16 Lift and drag on a rectangular building.

The lift F_L of any object can be calculated from

$$F_L = C_L A_o \frac{\rho V^2}{2} \qquad (4.22)$$

where C_L is the **lift coefficient** and A_o is a characteristic area that may or may not be the same as the characteristic area for drag.

For a structure that resembles a flat plate, such as a billboard or a stop sign, large lift can be generated at small angle of attacks, say, 10 degrees. Particularly large lift can be generated on cambered streamlined objects such as an airfoil at a small angle of attack. At a large angle of attack, flow separation takes place, causing a sudden reduction in lift called **stall**.

Unsteady lift. The alternate shedding of vortices from the two sides of a cylinder as in Figure 4.9(a) generates an alternating pressure across the cylinder. This causes a large dynamic (unsteady) lift on the cylinder which can cause the cylinder to vibrate laterally. Other objects exposed to wind also generate dynamic lift forces, even though they may not generate static (steady-state) lift.

Sec. 4.4 Wind Forces and Moments

The rms (root-mean-square) of the dynamic lift force on a structure can be calculated from

$$(F_L)_{rms} = (C'_L)_{rms} A_o \frac{\rho V^2}{2} \qquad (4.23)$$

where $(C'_L)_{rms}$ is the root-mean-square of the dynamic lift coefficient C'_L. Note that there is no relationship between the steady-lift coefficient C_L and the dynamic-lift coefficient C'_L. While C_L is zero for symmetric flow around a symmetric cylinder, C'_L is not zero due to vortex shedding.

More about unsteady lift will be discussed in the next chapter.

Normal Forces

For a rectangular structure such as the one shown in Figure 4.16, often it is of interest to determine the forces F_x and F_y normal to the walls. They can be calculated from the lift and the drag. From Figure 4.16,

$$F_x = F_D \cos \alpha - F_L \sin \alpha \qquad (4.24)$$

$$F_y = F_D \sin \alpha + F_L \cos \alpha \qquad (4.25)$$

Alternatively, F_x and F_y can be calculated from

$$F_x = C_x A_o \frac{\rho V^2}{2} \qquad (4.26)$$

$$F_y = C_y A_o \frac{\rho V^2}{2} \qquad (4.27)$$

where A_o is the area of the structure projected on a plane normal to the y-axis and C_x and C_y are force coefficients based on this projected area. The use of C_x and C_y instead of C_D and C_L in this case yields directly the forces along the two axes of the structure, x and y.

Vertical Overturning Moment

Structures must be designed to withstand the shear force developed in structures, F_S, and the vertical overturning moment, M, generated by drag and lift. In cases where lift is insignificant, the shear force is simply equal to the drag F_D, and the overturning moment M is F_D

multiplied by a moment arm $\bar{z}$. The moment arm is the distance above ground at which the force F_D must be applied to, namely,

$$M = \bar{z}\, F_D \tag{4.28}$$

In terms of the **moment coefficient**, C_M,

$$M = C_M A_o h \frac{\rho V^2}{2} \tag{4.29}$$

where h is a characteristic height of the building such as the building height H.

Example 4-3. Determine the moment coefficient C_M generated by wind on the telephone pole considered in the previous example. Suppose the line of action of the total drag on the pole is 18 ft above ground, and the characteristic length h used in Eq. 4.29 is the height of the pole.

[Solution]. From the calculations in the previous example, $F_D = 327$ lb. Then from Eq. 4.28, $M = 18 \times 327 = 5886$ ft-lb. Finally, from Eq. 4.29, $C_M = 2 \times 5886/(0.00237 \times 117.4^2 \times 30 \times 0.833 \times 30) = 0.48$.

Note that the moment coefficient C_M is normally determined from wind tunnel tests. Once C_M is known, the overturning moment can be calculated easily from Eq. 4.29.

Horizontal Twisting Moment (Torsion)

Some structures such as an L-shaped building may encounter a large horizontal twisting moment or torsion. They can be determined once the horizontal forces on all walls and their points of action are determined. Alternatively, an equation similar to Eq. 4.29 can be used to determine this moment once the coefficient of this moment is determined from a wind tunnel model test.

Shear (Fluid Friction) on Structures

In all the foregoing discussions, it has been assumed implicitly that the only forces acting on structures are pressure forces, that is, forces in the direction perpendicular to structural surfaces. In reality, structural surfaces are not only subjected to pressure forces but also to shear (fluid

Sec. 4.4   Wind Forces and Moments   105

friction) which is tangent to surfaces and in the direction of wind. However, since shear stresses generated by flows are normally two orders of magnitude smaller than pressure, the shear forces on structures, called "skin drag" or "frictional drag," can normally be neglected safely. In most cases, one need only consider the forces caused by pressure—called "pressure drag" or "form drag" in the along-wind direction and "lift" when perpendicular to wind. The only exception is in cases where a structural surface parallel to wind is very large, as for the case of a low-rise building with an unusually large depth D—the horizontal dimension parallel to wind. In such a case, the skin drag may be significant.

It is recommended here that whenever D/H or D/B is greater than 10, contribution to wind load by shear be considered. The 1989 Australian standard gives the following formulas for calculating the skin drag on rectangular buildings: For $H/B \leq 1$,

$$F_S = C_S B (D - 4H)q + C_S 2H(D - 4H)q \quad (4.30)$$

For $H/B > 1$,

$$F_S = C_S B (D - 4B)q + C_S 2H(D - 4B)q \quad (4.31)$$

where F_S is the total skin drag on the building; C_S is the skin drag coefficient; q is the dynamic pressure $\rho V^2/2$; and B, D, and H are the breadth, depth, and height of the building.

The Australian standard gives $C_S = 0.01$ for smooth surfaces with or without corrugations or ribs parallel to wind direction, $C_S = 0.02$ for surfaces with corrugations normal to wind direction, and $C_S = 0.04$ for surfaces with ribs normal to wind.

Note that the first term on the right side of Eqs. 4.30 and 4.31 gives the skin drag on the roof, and the second term in these equations gives the skin drag on the two side walls. The two terms are separated in the equations so that the skin drags on the side walls and on the roof can be caculated separately, if necessary.

Example 4-4. For a rectangular building of a dimension $H = 10$ m, $B = 80$ m, and $D = 100$ m, what is the total drag on the building including both form drag and skin drag? The wind speed at the roof level is 30 m/s, and the density of the air is 1.1 kg/m³.

[Solution]. Assume the surface of the building including both walls and the roof is smooth, meaning that $C_S = 0.01$. For air with density equal to 1.1 kg/m^3, the dynamic pressure is $q = 1.1 \times 30^2/2 = 495$ Pa. Because $H/B = 10/80 = 0.125 < 1$, Eq. 4.30 is applicable, and it yields a skin drag of $F_S = 0.01 \times 80 \times (100 - 4 \times 10) \times 495 + 0.01 \times 2 \times 10 \times (100 - 4 \times 10) \times 495 = 23,760 + 5,940 = 29,700$ N.

To calculate the form drag on the building, assume the pressure coefficient to be 0.8 for the windward wall and -0.5 for the leeward wall. This yields a form drag coefficient of $C_D = 0.8 - (-0.5) = 0.8 + 0.5 = 1.3$ From Eq. 4.21, the form drag is $F_D = 1.3 \times (10 \times 80) \times 1.1 \times 30^2/2 = 514,800$ N, which is more than 10 times greater than the skin drag. Finally, combination of the skin drag with the form drag yields a total drag of $F_D = 29,700 + 514,800 = 544,500$ N.

This example shows that even for a building with a depth (length in the wind direction) 10 times that of the building height, the skin drag was still one order of magnitude smaller than the form drag. This gives justification for neglecting skin drag when D/H or D/B is less than approximately 10.

REFERENCES

AKINS, R. E. AND CERMAK, J. E. (1975). *Wind Pressure on Buildings,* Technical Report CER76-77REA-JEC15, Fluid Dynamics and Diffusion Laboratory, Colorado State University, Fort Collins, Colorado.

ASCE (1961). "Wind Forces on Structures," *Transaction of American Society of Civil Engineers,* 126(2), 1124–1198.

ASCE (1987). *Wind Loading and Wind Induced Structural Response,* State-of-the-Art Report, Committee on Wind Effects, American Society of Civil Engineers, New York.

BEST, R. J. AND HOLMES, J. D. (1978). *Model Study of Wind Pressure on an Isolated Single-Story House.* Report 3/78, James Cook University, Townsville, North Queensland, Australia.

CERMAK, J. E. (1977). "Wind Tunnel Testing of Structures," *Journal of Engineering Mechanics Division,* ASCE, 103(6), 1125–1140.

DAVENPORT, A. G., SURRY, D. AND STATHOPOULOS, T. (1977). *Wind Loads on Low-Rise Buildings, Final Report on Phases I & II,* BLWT-SS8, University of Western Ontario, London, Ontario, Canada.

DAVENPORT, A. G., SURRY, D. AND STATHOPOULOS, T. (1978). *Wind Loads on Low-Rise Buildings, Final Report on Phase III,* BLWT-SS8, University of Western Ontario, London, Ontario, Canada.

JENSEN, M. AND FRANK, N. (1963). *Model-Scale Tests in Turbulent Wind.* Part I: *Phenomena Dependent on the Wind Speed,* Danish Technical Press, Copenhagen.

JENSEN, M. AND FRANK, N. (1965). *Model-Scale Tests in Turbulent Wind.* Part II: *Phenomena Dependent on the Velocity Pressure,* Danish Technical Press, Copenhagen.

LIU, H. (1975). "Wind Pressure Inside Buildings," *Proceedings of the 2nd U.S. National Conference on Wind Engineering Research,* Colorado State University, Fort Collins, III.3.1–3.3.

LIU, H. AND DARKOW, G. L. (1989). "Wind Effect on Measured Atmospheric Pressure," *Journal of Atmospheric and Oceanic Technology,* American Meteorological Society, 6(1), 5–12.

LIU, H. AND SAATHOFF, P. J. (1981). "Building Internal Pressure: Sudden Change," *Journal of Engineering Mechanics Division,* ASCE, 107(2), 309–321.

LIU, H. AND SAATHOFF, P. J. (1982). "Internal Pressure and Building Safety," *Journal of Structural Division,* ASCE, 108(10), 2223–2234.

PETERKA, J. A. AND CERMAK, J. E. (1974). "Wind Pressure on Buildings— Probability Densities," *Journal of Structural Division,* ASCE, 101(6), 1255–1267.

PROEPPER, H. AND WELSCH, J. (1979). "Wind Pressures on Cooling Tower Shells," *Proceedings of the 5th International Conference on Wind Engineering,* J. E. Cermak, editor, Pergamon Press, New York.

ROSHKO, A. (1961). "Experiments on the Flow Past a Circular Cylinder at Very High Reynolds Number," *Journal of Fluid Mechanics,* 10, 345–356.

SIMIU, E. AND SCANLAN, R. H. (1986). *Wind Effects on Structures* (2nd ed.), John Wiley & Sons, New York.

STATHOPOULOS, T. AND BASKARAN, A. (1985). "The Effect of Parapets on Wind-Induced Roof Pressure Coefficients," *Proceedings of the 5th U.S. National Conference on Wind Engineering,* Texas Tech University, Lubbock, 3A.29–36.

5

Dynamic Response of Structures: Vibration and Fatigue

5.1 INTRODUCTION

Wind-Sensitive Structures

Many structures, such as tall buildings, stacks (chimneys), towers, cable-suspended bridges, cable-suspended roofs, light poles, traffic light fixtures, signs, sheet metal roofs, and so on, are sensitive to wind-induced vibration and resultant damage. The trend toward constructing higher buildings and longer bridges with less material has contributed to a new generation of wind-sensitive structures that the modern-day structural engineer must cope with.

Factors Contributing to Wind-Induced Vibration

Wind-induced structural vibration depends to a large extent on the characteristics of structures. The three most relevant characteristics are shape, stiffness (flexibility), and damping.

Wind-induced dynamic forces that cause a structure to vibrate depend on the shape of the structure. For instance, long cylinders (especially circular cylinders) exposed to winds normal to the cylinders create vortex shedding, which in turn can cause large vibrations. Slender cross sections (plates) such as a stop sign are susceptible to flutter—a special type of vibration that will be discussed later. As to the effect of stiffness, the stiffer (less flexible) a structure is, the less susceptible it is to wind-induced vibration. For instance, while a large panel of sheet metal may vibrate violently in wind, increasing its stiffness by adding braces to the panel will eliminate or greatly reduce vibration. Finally, damping reduces vibration. Damping can be caused by mechanical friction in materials or between structural parts, by the interaction of a structure with its foundation, by the mass of the structure itself, or by external devices such as shock absorbers.

In addition to the aforementioned structural characteristics, wind-induced structural vibration also depends on the characteristics of wind. For instance, the turbulence in wind buffeting on structures causes vibration. Large integral scale and high turbulence intensity in the free stream cause strong buffeting. Buffeting can be especially serious if the dominant frequency of turbulence as indicated by the spectral peak (see Figure 3.5) approaches the natural frequency of the structure.

In the case of vortex shedding from two-dimensional structures (long cylinders), the effect of turbulence is the opposite. Smooth, uniform wind (i.e., a wind with little turbulence and a uniform velocity profile) produces more regular vortex shedding than a turbulent wind with spanwise velocity variation.

Aerodynamic and Aeroelastic Instabilities

Aerodynamic instability or simply **aeroinstability** is the instability (variation or fluctuations) of the air flow around structures, such as the wake generated downwind of bluff bodies and the vortex shedding created by cylinders. When a structure encounters aeroinstability, it may vibrate or suddenly deflect in the air flow. The structural motion in turn modifies the flow around the structure. This interaction between structural motion and wind is called **aeroelasticity** or **aeroelastic phenomena**. If the modification of wind pattern around a structure by aeroelasticity is such that it increases rather than decreases vibration or deflection, the phenomenon is called **aeroelastic instability**, or **negative damping**. Many wind-induced vibrations are caused by aeroelastic instability.

5.2 TYPES OF STRUCTURAL VIBRATION

Depending on its cause, wind-induced structural vibration can be classified into many types, as discussed in the paragraphs that follow.

Vortex-Shedding Vibration

The flow behind a long cylinder held perpendicular to wind is characterized by the periodic shedding of vortices briefly described in the previous chapter. Vortex shedding creates periodic lateral forces that can cause vibration of slender structures such as stacks, masts, towers, and tall buildings. Although vortex shedding is most noticeable for circular cylinders, it also happens to a lesser degree to cylinders of any other shape.

A dimensionless parameter commonly used to study the frequency of vortex shedding is the Strouhal number defined as

$$S = \frac{nD}{V} \quad (5.1)$$

where n is the shedding frequency in hertz, D is the diameter of the cylinder, and V is the wind speed relative to the cylinder.

For noncircular cylinders, the Strouhal number is defined the same as in Eq. 5.1 except that a characteristic length, other than the diameter, is used. Normally, the characteristic length used is the dimension of the cross section in the across-wind direction—the breadth B.

The Strouhal number is a function of the Reynolds number. As shown in Figure 5.1, the Strouhal number of a long circular cylinder in a smooth uniform flow is approximately 0.21 over the Reynolds number range $300 < R < 2 \times 10^5$—the subcritical range. The Reynolds number 2×10^5 is the *critical Reynolds number* that produces a sudden shift of the separation point and a sudden decrease in the drag coefficient; see Figure 4.14. As described in Chapter 4, the subcritical range is characterized by regular (periodic) shedding of vortices. Within this range, the shedding frequency of a long circular cylinder may be predicted approximately from

$$n = 0.21 \frac{V}{D} \quad (5.2)$$

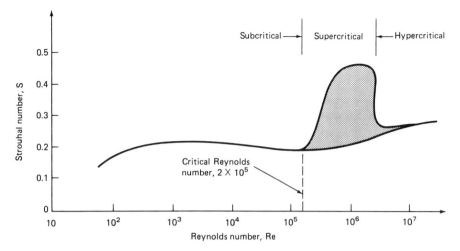

Figure 5.1 Variation of Stouhal number with Reynolds number for circular cylinder, smooth flow.

Note that with turbulence in the free stream, the value of 0.21 is slightly decreased to approximately 0.20.

For Reynolds number between 2×10^5 and 4×10^6 (the *supercritical range*), vortex shedding becomes more random, and the Strouhal number, based on the dominant frequency, increases to about 0.5. For Reynolds number greater than 4×10^6 (the *hypercritical range*), the vortex shedding again becomes more regular (periodic), and the Strouhal number becomes approximately 0.25.

Note that the frequency n in Eqs. 5.1 and 5.2 represents the number of vortices shed per second from one side of a cylinder. The number of vortices shed per second from both sides of a cylinder is $2n$. Since it takes the shedding of two vortices (one from each side) to complete one cycle of oscillation of the lateral force by vortex shedding, the frequency of the lateral-force oscillation caused by vortex shedding is n. In contrast, a structure downstream from a cylinder will encounter two vortices per cycle of shedding. Thus, the frequency of the dynamic force hitting the downstream structure is $2n$.

Vortex shedding generated vibration takes place when the wind speed is such that the shedding frequency becomes approximately equal to the natural frequency of the cylinder—a condition that causes *resonance*. When resonance takes place, further increase in wind speed by a few percent will not alter the shedding frequency. The shedding is now

Sec. 5.2   Types of Structural Vibration   113

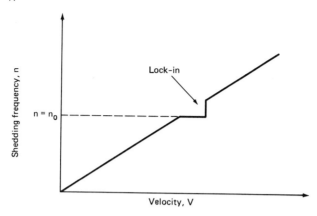

Figure 5.2   Lock-in phenomenon.

controlled by the natural frequency of the structure, n_o. This phenomenon, illustrated in Figure 5.2, is called **lock-in**. Finally, when wind speed is increased to above that causing lock-in, the frequency of shedding will again be controlled by the wind rather than the structural frequency. Because the structure vibrates excessively only in the lock-in range, having a wind speed either below or above the lock-in range will not cause serious vibration except in the second or higher modes.

Note that the shedding frequency given by Eq. 5.2 and Figure 5.1 is that for a slender circular cylinder in a uniform flow containing little turbulence—results obtained in aeronautical wind tunnels which have uniform velocity and less than 1% relative intensity of turbulence. Because natural winds have strong turbulence and variation of wind speed with height, Eq. 5.2 and Figure 5.1 can be used as an approximation for natural wind-induced vortex shedding only if the variation of the wind velocity over the length of a cylinder is negligible, as for the case of a horizontal cylinder, and if the gust speed instead of the mean wind speed is used to calculate the Strouhal number in a quasi-steady approach—a valid approach when the cylinder diameter is much smaller than the average size of the eddies in the wind.

For tall vertical cylinders such as a chimney exposed to natural winds, due to the variation of the mean wind speed with height, Eq. 5.2 and Figure 5.1 must be modified. In such a case, if the velocity V in the Strouhal number represents the hourly mean velocity at the top of the cylinder, the shedding frequency can be predicted approximately as follows (NBC Supplement 1985):

$$n = \frac{V}{6D} \quad \text{if } nD^2 < 0.5 \text{ m}^2/\text{s}$$

$$n = \frac{1}{3D}(V - \frac{1.5}{D}) \quad \text{if } 0.5 \text{ m}^2/\text{s} < nD^2 < 0.75 \text{ m}^2/\text{s} \quad (5.3)$$

$$n = \frac{V}{5D} \quad \text{if } nD^2 > 0.75 \text{ m}^2/\text{s}$$

where D must be in meter and V in meters per second.

Example 5-1. Suppose the telephone pole discussed in Example 4-2 has a first-mode structural frequency of 10 hertz. At what wind speed will the pole vibrate dangerously due to lock-in? What would happen if the structural frequency were much higher or lower?

[Solution] With $n = 10$ Hz and $D = 10$ in. $= 0.245$ m, we have $nD^2 = 0.645$ m^2/s. From the second expression in Eq. 5.3, $V = 13.53$ m/s $= 30.3$ mph. Since this is an hourly mean wind speed that may occur more than once each year (return period less than a year), and since this wind speed can cause moderate forces, the lock-in is of concern in this case. Frequent vibration of this pole in the neighborhood of 30 mph may cause fatigue damage to the pole.

If a pole with a higher structural frequency, say, 100 hertz, is used, it yields a wind speed of 127 m/s (284 mph)—a speed that is unlikely to occur within the life of the structure. This is safe for the pole. On the other hand, if a pole with a much lower frequency, say, 1 hertz, is used, the wind speed corresponding to lock-in will be 1.52 m/s (3.4 mph). Although this is a rather small and hence safe wind speed, vibration at higher modes, corresponding to higher speeds, is possible.

Vortex shedding is also generated on cylinders having other cross-sectional geometries such as H and L shapes. The Strouhal number for those cylinders in smooth, uniform flow is often in the range of 0.12–0.16. A list of the Strouhal number for cylinders of various cross-sectional shapes can be found in Simiu and Scanlan (1986) and Blevins (1977). These two references also contain a wealth of information on structural vibration of various kinds.

A dynamic lift is generated by vortex shedding. This force per unit length of a cylinder varies with time approximately in a sinusoidal manner as follows:

$$f_L = \frac{1}{2} \rho V^2 D\, C'_L \sin(2\pi n t) \quad (5.4)$$

where f_L is the lift force per unit length of the cylinder, D is the diameter of the cylinder (use width B if the cylinder is noncircular), C_L' is the dynamic lift coefficient, n is the shedding frequency determined from Eq. 5.1, and t is time. As explained in Chapter 4, the dynamic lift coefficient C_L' is quite different from the steady lift coefficient C_L. For a circular cylinder in the subcritical range, C_L' is approximately equal to 0.6.

A problem in using Eq. 5.4 is the imperfect correlation of lift along any cylinder. Due to this incoherence, one cannot simply multiply the force calculated from Eq. 5.4 by the length of the cylinder to determine the total lift on the cylinder. Besides, as soon as a cylinder begins to vibrate in wind, aeroelastic effects set in, causing the lift coefficient C_L' to increase with increasing amplitude of oscillation. The spanwise correlation also increases with the oscillation amplitude. Furthermore, free-stream turbulence and spanwise variation of local mean velocity both reduce vortex shedding drastically. Due to these complexities, the dynamic lift cannot be predicted or prescribed with accuracy in any real case.

Many past studies dealt with finding a suitable expression for f_L to fit experimental vibration data. A good summary of these studies is contained in Simiu and Scanlan (1986). Generally, f_L is considered to vary not only with time t but also with the amplitude of vibration X and its first and second derivatives, $\dot{X}$ and $\ddot{X}$. Once the functional form of f_L is determined, Eq. A.20 or A.21 can be used to analyze the vibration of the cylinder.

Vibration at Critical Reynolds Number

While vortex shedding causes lateral vibration of cylinders, the rather sudden drop of the drag coefficient at the critical Reynolds number (2×10^5) can create a longitudinal (along-wind) vibration. When the mean wind speed is such that the Reynolds number is in the neighborhood of 2×10^5, large fluctuations of the wind speed about this mean value cause the cylinder to fluctuate between the subcritical and the supercritical range. Since the drag coefficient is quite different for these two ranges, large vibration can result in the along-wind direction.

Across-Wind Galloping

Under the action of steady wind, ice-coated transmission lines, cable-suspended bridge decks, and some other flexible structures are often plagued by low-frequency, large-amplitude, across-wind vibration called **galloping**. At moderate wind speeds, galloping of ice-coated transmission lines of an amplitude greater than 5 m (16 ft) is not unusual. Galloping can happen to almost any lightweight, flexible, cylindrical (prismatic) structures, except those of circular cross section, exposed to wind. However, some special cross-sectional shapes, such as rectangular sections or D sections, are more prone to galloping than others.

In the analysis of galloping, it is assumed that the vibration is slow so that a quasi-steady approach is valid for determining the lift and drag from the following steady-state equations:

$$f_L = C_L B \frac{\rho V_r^2}{2} \quad (5.5)$$

and

$$f_D = C_D B \frac{\rho V_r^2}{2} \quad (5.6)$$

where f_L and f_D are the lift and drag per unit length of the cylinder; C_L and C_D are the steady-state lift and drag coefficients that are functions of α, the angle of attack of wind with respect to the cylinder; B is the breadth (width) of the cylinder; ρ is the density of the air; and V_r is the velocity of wind relative to the cylinder. Note that it is always the relative velocity between structure and wind that creates the wind pressure, drag, and lift on the structure.

Galloping is an instability that must be initiated by a disturbance in wind (turbulence component in the across-wind direction) that causes a crosswise vibration. As a cylinder vibrates crosswise in a steady wind of velocity V, both the magnitude and the direction of the relative velocity of V_r change, which in turn causes the angle of attack α to change. For instance, as shown in Figure 5.3, a downward movement of the cylinder at speed dy/dt or $\dot{y}$ will cause an increase in the angle of attack by an amount equal to $\dot{y}/V$. Depending on the cross-sectional shape of the cylinder and the angle of attack, the increase in α will either increase or decrease the lift on the cylinder. If the lift is increased by the increase in α, the motion of the cylinder in the y-direction is inhibited by the lift increase and hence the situation is inherently stable; no large vibration can occur. On the other hand, if the increase in α causes a decrease in lift in the opposite direction of vibration, then the situation is unstable and conducive to galloping.

Sec. 5.2 Types of Structural Vibration

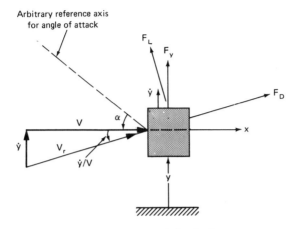

Figure 5.3 Across-wind galloping.

More specifically, galloping occurs when the y-component of this lift decrease is greater than the y-component of drag increase. A necessary condition of this unstable situation (galloping) is

$$\frac{dC_L}{d\alpha} + C_D < 0 \tag{5.7}$$

which is called the **Glauert-Hartog criterion**. From Eq. 5.7, a structure may gallop only if the vertical force coefficient, C_L, when plotted as a function of α, exhibits a negative slope having an absolute value greater than the drag coefficient C_D at the same α value.

A more exact condition (sufficient condition) of galloping is

$$\frac{dC_L}{d\alpha} + C_D < -\frac{2c}{\rho V B} \tag{5.8}$$

where c is the damping constant used in the study of across-wind vibration of a cylinder; see Eq. A.20.

Since the right side of Eq. 5.8 is negative, the left side must be more negative (i.e., having a larger absolute value) than the right side for the inequality to hold. This means the quantity $dC_L/d\alpha$ must be even more negative than required by Eq. 5.7 before galloping will take place.

Due to its negative sign, the right side of Eq. 5.8 is larger (closer to zero) when the velocity V is higher. Therefore, increasing wind velocity increases the tendency of galloping. Because the right side of Eq. 5.8 goes to negative infinity when V is zero, galloping will not occur unless the wind speed exceeds a certain minimum value. The wind speed required to cause a flexible structure to gallop is small as compared to that of causing a rigid structure to gallop. Usually, galloping has much lower frequency and much larger amplitude than vibration by vortex shedding from the same structures. Structures gallop at their natural frequencies.

Although Eq. 5.8 is more accurate than Eq. 5.7, the latter is used more often because it is simpler and more conservative for predicting galloping. The variation of C_L with α required for determining the quantity $dC_L/d\alpha$ is normally obtained from wind tunnel tests.

Example 5-2. The lift and draft coefficients for an oxagonal cylinder are given as a function of the angle of attack α in Figure 5.4. Determine whether and when the cylinder will gallop.

[Solution]. From Figure 5.4, there are two places where the lift coefficient has a negative slope: where α is between -5 and 5 degrees and where α is between 40 and 45 degrees. The former represents a wind direction within a 5-degree angle each way perpendicular to the side shown in the sketch; the latter represents, within experimental error, a similar wind direction perpendicular to a neighboring side of the oxagon. Since the lift and drag coefficients for these two analogous places are slightly different due to experimental error, they are separately investigated for possible galloping as follows:

In the range $-5° < \alpha < 5°$, the slope of the C_L-curve is approximately $dC_L/d\alpha \approx -0.45/4° = -0.113/\text{degree} = -6.4/\text{radian}$, and $C_D = 1.0$. Thus, $dC_L/d\alpha + C_D \approx -6.4 + 1.0 = -5.4 < 0$. Since the Glauert-Den Hartog criterion is satisfied in this region, galloping may occur here.

In the range $40° < \alpha < 45°$, $dC_L/d\alpha \approx -0.43/2.7° = -0.159/\text{degree} = -9.1/\text{radian}$, and $C_D = 0.8$. Thus, $dC_L/d\alpha + C_D \approx -9.1 + 0.8 = -8.3 < 0$. Again the Glauert-Den Hartog criterion is satisfied here.

The foregoing calculation shows that galloping may occur to the oxagonal cylinder in the neighborhood of α equal to 0 and 45 degrees, that is, when the wind is perpendicular or nearly perpendicular to any side of the oxagon. Note that fulfillment of the Glauert-Den Hartog criterion does not guarantee that galloping will take place; it merely establishes the possibility of galloping. On the other hand, when the Glauert-Den Hartog criterion is not satisfied, the structure definitely will *not* gallop.

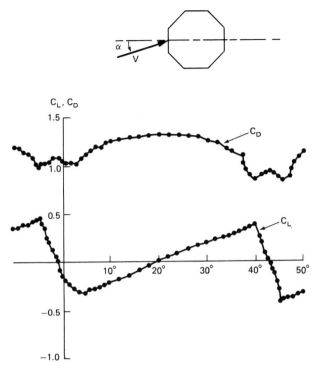

Figure 5.4 Drag and lift coefficients for octagonal cylinder at $Re = 1.2 \times 10^6$. (From Scanlan and Wardlaw, 1973.)

Wake Galloping

Consider two cylinders separated at a few diameters away from each other with one cylinder in the wake of the other; see Figure 5.5. Due to the circulation of the flow inside the wake (clockwise for the upper half of the wake in Figure 5.5 and counterclockwise for the lower half), the cylinder located in the upper half of the wake, if allowed to move, will oscillate in a clockwise, elliptic path as shown. Likewise, a cylinder free to move in the lower half would oscillate in the counterclockwise direction. Such oscillation or vibration is called **wake galloping**. It is a phenomenon often observed in power transmission lines held in "bundles" by spacers. With spacers separated at large distances apart along power lines, the lines between two neighboring spacers can move laterally for large distances and cause large wake galloping. Decreasing the distance between spacers and increasing the tension of the lines all tend to reduce wake galloping.

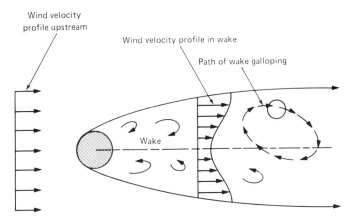

Figure 5.5 Wake galloping.

Torsional Divergence

Torsional divergence is a phenomenon discovered in aeronautics. It refers to the large twisting (torsional) moments generated on aircraft wings which may twist the wings off aircraft flying at high speeds. Thin plate structures exposed to wind, such as the deck of a cable-suspended bridge, may also be affected by torsional divergence at high wind velocity.

The mechanism of torsional divergence is the same as for galloping, except that torsion (twisting) instead of linear vibration is involved. The way torsional divergence is generated can be explained by considering a thin airfoil (i.e., the cross section of an aircraft wing) or a thin bridge deck as shown in Figure 5.6. At small angle of attack (α between 0 and 20 degrees), not only is a drag F_D and a lift F_L generated, a clockwise moment M is also produced on the airfoil and the bridge deck. This moment causes the angle α to increase, which in turn causes a larger twisting moment that further increases α. If the airfoil does not have sufficient torsional stiffness to resist this increasing moment, the condition is unstable and the foil will be twisted to failure.

Torsional divergence can be analyzed by considering the structure (airfoil on bridge deck) being supported by a torsional spring at the elastic center of the structure as shown in Figure 5.6. The aerodynamic moment M per unit span length generated by wind speed V relative to the structure is

$$M = \frac{1}{2}\rho V^2 B^2 C_M(\alpha) \qquad (5.9)$$

Sec. 5.2 Types of Structural Vibration

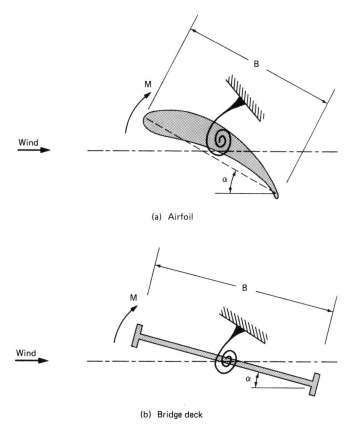

Figure 5.6 Analysis of torsional divergence.

where B is the width of the bridge deck or the chord of the airfoil and C_M is the moment coefficient which is a function of α.

Under equilibrium, the aerodynamic moment is balanced by the resisting moment $k_T \alpha$, where k_T is the torsional spring constant, namely,

$$\frac{1}{2} \rho V^2 B^2 C_M(\alpha) = k_T \alpha \qquad (5.10)$$

Torsional divergence takes place when the left side of Eq. 5.10 is greater than the right side. This leads to the following criteria for initiation of torsional divergence:

$$\frac{dC_M}{d\alpha} > \frac{2k_T}{\rho V^2 B^2} \qquad (5.11)$$

Because the right side of Eq. 5.11 is positive, a necessary condition for torsional divergence is $dC_M/d\alpha > 0$, a condition similar to the Glauert-Den Hartog criterion for across-wind galloping. This means that a curve giving the variation of C_M with α can be used to assess the possibility of torsional divergence in the same manner the C_L and C_D curves are used for assessing across-wind galloping.

The minimum velocity required to initiate torsional divergence is called the *critical divergence velocity*, V_c. From Eq. 5.10, it can be proved that (Simiu and Scanlan 1986)

$$V_c = \sqrt{\frac{2k_T}{\rho B^2 C_M'}} \qquad (5.12)$$

where C_M' is the derivative $dC_M/d\alpha$. The value of C_M' can be determined from a graph of C_M versus α obtained through wind tunnel tests.

Flutter

The term **flutter** has been used by different investigators to describe different types of wind-induced vibration of thin plates or airfoils. Four kinds of flutter are discussed next.

Classical flutter. The classical flutter or simply **flutter** is a two-degree-of-freedom vibration involving simultaneous lateral (across-wind translational) and torsional (rotational) vibrations. It occurs in structures that have approximately the same magnitude of natural frequencies for both the translational and the rotational modes. Similar to galloping and torsional divergence, flutter is produced by aerodynamic instability completely unrelated to vortex shedding. Unlike galloping and torsional divergence, the two-degrees-of-freedom of a structure— rotation and translation—are coupled together and hence affect each other in flutter. The lift and drag coefficients used in the analysis of flutter are not only functions of α but also functions of $\dot{\alpha}$ (angular speed of flutter), $\dot{h}$ (lateral speed of flutter), and the reduced frequency $B\omega/V$, where ω is the angular frequency of flutters ($\omega = 2\pi n$). Flutter is a phenomenon that had been explored thoroughly in aeronautics in association with the flutter of aircraft wings, before the concept was used in bridge design.

Stall flutter. Stall is the sudden drop of lift and the accompanying increase in drag on an airfoil or flat plate at large angles of attack. It is a

Sec. 5.2 Types of Structural Vibration

rather unstable condition that causes aircraft to crash and certain ground-based structures such as traffic signs to vibrate violently—the **stall flutter**. Stall flutter of signs occurs when the wind is oblique to the sign at an angle of attack near where stall (flow separation) is initiated. With the sign mounted on a central post as in the case of a stop sign, in moderate to high winds the sign vibrates about the axis of the post in violent twists. Such vibration can be reduced by using sign posts of greater torsional stiffness or using greater damping.

Panel flutter. The surface panel of a rocket flying at supersonic speed may vibrate in a mode called **panel flutter**. This type of vibration also can happen at much lower velocities (less than 20 m/s) if the panel is somewhat flexible, such as an air-lift dome, a canvas cover of a boat, or a sheet metal roof that has not been properly fastened. Flag flutter is a close relative of the panel flutter.

Single-degree-of-freedom flutter. This includes stall flutter and other single-degree-of-freedom flutters, either linear or torsional. For instance, suspended bridge decks can be excited by flow separation to cause the decks to undergo single-degree-of-freedom torsional movements.

Buffeting Vibration

All the types of vibration discussed before except for wake galloping are self-induced by structures. They exist even when the approaching free flow is without turbulence—in fact, they are often stronger in smooth flow than in turbulent flow. In contrast, buffeting vibration is the vibration produced by turbulence or other disturbances of the flow not generated by the vibrating object itself. Two types of buffeting exist: those caused by free-stream turbulence and those caused by disturbances generated by an upwind neighboring structure or obstacle. The latter type is called **wake buffeting** or **interference.**

The most typical interference is between neighboring skyscrapers in urban areas, such as the twin towers of the World Trade Center in New York City. The turbulence of the upstream building buffeting on the downstream building causes the latter to vibrate. Other neighboring structures, such as large chimneys, may also cause serious buffeting vibration to buildings. For instance, the building shown in Figure 5.7 is buffeted by the vortex shedding of the upstream cylinder—a chimney. The building is hit by regular vortices at a frequency twice that predicted

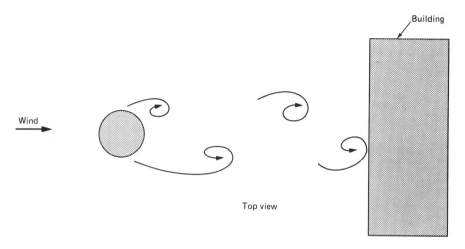

Figure 5.7 Buffeting on building by wakes generated by upwind structures.

from Eq. 5.3. Keep in mind, however, that the vortices generated on the two sides of the chimney do not hit the same place on the building's windward wall. The spacing between the two rows of vortices may cause a horizontal twisting moment on the building having a rotational frequency n predicted from Eq. 5.3.

Because turbulence is three dimensional, buffeting by turbulence (both free-stream turbulence and wake turbulence) can produce not only along-wind vibration but also across-wind vibration and even twisting (torsional vibration).

Wind-Induced Building Vibration

Wind-induced vibration of buildings is often a combination of the various types of vibrations discussed before, with turbulence always playing a role. Because the energy of turbulence is concentrated in the low-frequency range of the spectrum, building vibration caused by turbulence is important only when the natural frequency of the building is low (say, less than 1 hertz), and when the damping of the building is light. More about building vibration will be discussed in Section 5.4.

5.3 MEANS TO REDUCE WIND-INDUCED VIBRATION

Many means are available to reduce wind-induced vibration on structures. These include (1) using lateral bracing (such as adding diagonals to trusses) to increase the stiffness of structures to resist wind forces and to

Sec. 5.3   Means to Reduce Wind-Induced Vibration   125

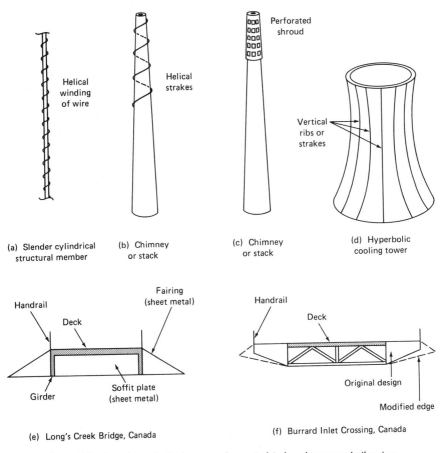

Figure 5.8 Aerodynamic devices to reduce wind-induced structural vibration.

reduce dangerous vibration, (2) using helical roughness around circular cylinders to reduce vibration due to vortex shedding (see Figure 5.8(a)), (3) modifying the upper portion of a chimney or stack by adding either strakes (as shown in Figure 5.8(b)) or a perforated shroud (as shown in Figure 5.8(c)) to reduce vibration due to vortex shedding, (4) using vertical ribs or strakes around cooling towers to reduce peak negative pressure and the vibration caused by it (Figure 5.8(d)), (5) modifying bridge shape to increase aerodynamic stability (Figures 5.8(e) and (f)), and (6) using mechanical dampers to reduce structure vibration.

Devices such as helical wire wrapped around a circular cylinder or strakes welded around a steel stack effectively reduce vibration by breaking up the regular pattern of vortices shed from the cylinder or stack, thereby making vortex shedding more random and less coherent. Such devices are called **vortex spoilers** or simply **spoilers**.

Wind-induced vibration of cable-suspended bridges can be reduced by adding fairings to bridge decks as shown in Figures 5.8(e) and (f). Studies have shown that suspended bridge decks of the open-box type can generate large vortex-shedding-induced vibration. However, by adding a simple triangular fairing on each side and a soffit plate on the bottom as shown in Figure 5.8(e), vertical vibration of the deck can be reduced by a factor of 10. The more complex shape of fairings in Figure 5.8(f) are even more effective.

Vibrations of tall buildings are often controlled by using one of two types of dampers: **tuned-mass dampers** and **viscoelastic dampers**. The tuned-mass damper (TMD) is a device consisting of a large mass, such as a huge concrete block, placed on the roof or an upper floor. The mass is connected to the building through a set of pneumatic springs (pistons) and a set of dashpots (hydraulic shock absorbers). The pneumatic springs are tuned to the actual frequency of the building as determined in the field. As the building vibrates in wind, the concrete block of the TMD, being supported by a thin film of oil on the floor, tends to counter the building vibration. By having the concrete block tuned to vibrate at the natural frequency of the building, the TMD will be most effective in countering the building motion. The system has fail-free devices to prevent excessive travel of the concrete block.

Examples of major buildings that have used TMD systems include the Citicorp Center in New York City and the John Hancock Building in Boston. A shortcoming of the TMD system is that it requires the use of electrical power to monitor the system behavior and activate the oil support system. The TMD does not function during power outages.

The viscoelastic dampers are a set of shock absorbers connected at various joints of a building. The World Trade Center uses approximately 10,000 viscoelastic dampers in each of the twin buildings. Power outages do not affect the operation of viscoelastic dampers.

5.4 PREDICTING DYNAMIC RESPONSE OF STRUCTURES

General Approaches

Wind-induced vibration to structures can be predicted directly from Eq. A.19 for a point-mass structure, from Eq. A .21 for a cylindrical structure, or from an extended form of these equations for other more complicated structures. Three common approaches for predicting structural vibration are the *time-domain approach,* the *frequency-domain approach,* and the *equivalent static-load approach.*

Time-domain approach. The time-domain approach solves Eq. A.19 or A.21 or an extended form of the equations directly. To do that, the forcing function $F(t)$ or $f(t)$, that is, the history of the load, must be known. The second-order linear differential equation is then solved numerically on a computer to yield the response X as a function of time t. Once $X(t)$ is known, one can easily determine the statistical properties of the response such as the root-mean-square, the peak, and the spectrum of X. The velocity and the acceleration of the structure can also be obtained, respectively, from the first and the second derivatives of X. The difficulty in using this method lies in that the forcing function $F(t)$ or $f(t)$ is not known normally.

Frequency-domain approach. In the frequency domain approach, the functions $F(t)$ or $f(t)$ are not needed; only their spectrum has to be known. Using the spectrum of $F(t)$, the spectrum of the response $X(t)$ can be determined easily from Eq. A.18. Of course, the mechanical admittance function H_1 that transforms the load spectrum to the response spectrum must also be known. From Eq. A.16, H_1 can be determined once the natural frequency n_o and the damping ratio are found. Once the spectrum of the response is known, the total root-mean-square of the response can be calculated from the area under the spectrum curve, namely,

$$X_{rms}^2 = \int_0^\infty S_X(n)\, dn \qquad (5.13)$$

Equivalent-static-load approach. An equivalent static load is the static load that would produce the same maximum stress and maximum deflection of the structure as those caused by the dynamic effect of gusts.

Suppose the mean, rms, and peak deflection of a structure at height z are, respectively, $\overline{X}$, X_{rms}, and $\hat{X}$. A peak factor, g, can be defined such that

$$\hat{X}(z) - \overline{X}(z) = gX_{rms}(z) \qquad (5.14)$$

With the use of 1 hour as the averaging time for $\overline{X}$ and X_{rms}, the value of g for tall structures is normally in the neighborhood of 3 to 4. Smaller values of g result if the mean wind speed is based on the fastest-mile wind.

Rearranging Eq. 5.14 yields

$$X(z) = \left(1 + \frac{gX_{rms}}{\overline{X}}\right)\overline{X} \qquad (5.15)$$

The quantity in the parentheses of the foregoing equation is called the **gust effect factor** in Canada, where the wind speed used for design is based on the hourly mean, and it is called the **gust response factor** in the United States, where the mean wind speed is the fastest-mile wind. In Europe and Australia, it is simply referred to as the **gust factor**. For simplicity and generality, we shall hereafter call it the gust factor.

From the foregoing, the gust factor is

$$G = 1 + g\frac{X_{rms}}{\overline{X}} \qquad (5.16)$$

Equation 5.16 gives the relationship between the gust factor G and the peak factor g. When the mean value is based on the fastest hourly wind speed, the value of G for tall structures is normally in the range of 2 to 3. Smaller values of G result when wind speeds are based on the fastest-mile wind speed.

Equation 5.15 can be rewritten as

$$\hat{X} = G\overline{X} \qquad (5.17)$$

This equation shows that the peak deflection $\hat{X}$ can be determined easily from the mean deflection, $\overline{X}$, once the gust factor G is known. Within the elastic range of structural materials, deflection is linearly proportional to load. Therefore, we may conclude from Eq. 5.17 that the peak load $\hat{F}$ responsible for producing $\hat{X}$ must be equal to G times the mean load $\overline{F}$ responsible for producing $\overline{X}$, namely,

$$\hat{F} = G\overline{F} \qquad (5.18)$$

Equation 5.18 shows that once the value of G for a given case is determined, the equivalent static load to produce $\hat{X}$ can be determined easily from the mean load.

To determine the gust factor G from Eq. 5.16, we must have a way to determine g, X_{rms}, and $\overline{X}$. The peak factor g can be determined from the following equation (Davenport 1967):

Sec. 5.4 Predicting Dynamic Response of Structures

$$g = (2 \ln RT)^{1/2} + \frac{0.577}{(2 \ln RT)^{1/2}} \qquad (5.19)$$

in which

$$R = \left(\frac{\int_0^\infty S_X(n) \, n^2 \, dn}{\int_0^\infty S_X(n) \, dn} \right)^{1/2} \qquad (5.20)$$

where T is the averaging time and R, according to Eq. 5.20, represents the distance (in frequency domain) from the center of the moment of inertia of the spectral graph S_X to the S_X-axis; see Figure 3.5. The value of R is directly proportional to, and somewhat larger than, the frequency of the spectral peak. When S_X has a dominating peak produced by resonance, the value of R approaches the frequency of the peak of this response spectrum. Thus, R will be referred to hereafter as the *dominant response frequency*. Eq. 5.19 can be represented graphically in Figure 5.9. Note that Eq. 5.19 does not hold for values of RT less than 2, approximately.

The quantity X_{rms} in Eq. 5.16 can be obtained from the Spectrum S_X and Eq. 5.13. The quantity $\overline{X}$ can be obtained from the calculated static deflection caused by the mean wind speed V. Using such an approach, Eq. 5.16 yields the gust factor needed for calculating the equivalent static load. To facilitate the use of this approach by designers, several investigators, including Davenport (1967), Vellozzi and Cohen (1968), and Simiu (1973, 1980), have prepared a set of graphs, tables, and simple formulas for determining G. Results of these three studies were used respectively in the National Building Code of Canada (1985), ANSI A58.1 (1972), and ANSI A58.1 (1982). These three methods may yield values of G different by more than 50%. Note that correct comparison of the G values between Canadian methods and U.S. methods must take into account the fact that the basic wind speed in the U.S. method is the fastest-mile speed, whereas Canadian calculations are based on the mean hourly fastest speed. The former is substantially greater than the latter; see Chapter 3.

Vibration of Cantilever Beam

A structure such as a tall building or a chimney is similar to a vertical cantilever beam. Therefore, it is of interest to discuss the vibra-

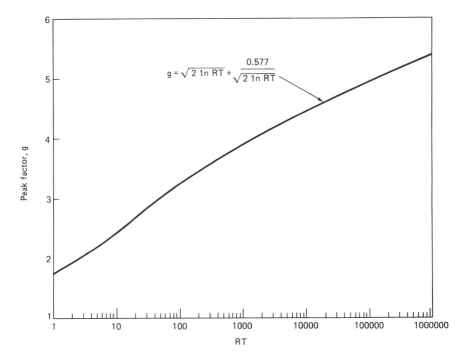

Figure 5.9 Variation of peak factor g with RT.

tion of cantilever beams before discussing the dynamic response of such structures.

The vibration of a cantilever beam is complicated by the fact that the beam may undergo simultaneous vibration in more than one mode. The first three modes of such vibration are illustrated in Figure 5.10.

If the beam or the beamlike structure is rather rigid, the beam vibrates mainly in its first mode, and Eq. A.19 reduces to

$$\ddot{X} + 2\omega_1 \zeta \dot{X} + \omega_1^2 X = \frac{P_1(t)}{M_1} \quad (5.21)$$

where X represents the displacement of the top of the beam, ω_1 and ζ are, respectively, the angular frequency and the damping ratio of the first mode of vibration, and $P_1(t)$ and M_1 are, respectively, the **generalized force** and the **modal mass** defined as follows:

$$P_1(t) = \frac{1}{H} \int_0^H p(z, t) z \, dz \quad (5.22)$$

Sec. 5.4 Predicting Dynamic Response of Structures

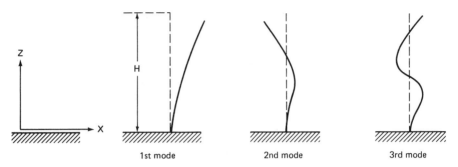

Figure 5.10 First three modes of vibration of a flexible cantilever beam.

and

$$M_1 = \frac{1}{H^2} \int_0^H z^2 m(z) \, dz \tag{5.23}$$

in which H is the height (length) of the beam, $p(z, t)$ is the distribution of load as a function of z and t, and $m(z)$ is the mass per unit length of the beam at z.

For a beam of uniform mass distribution along its length, m is constant, and Eq. 5.23 reduces to

$$M_1 = \frac{mH}{3} = \frac{M_b}{3} \tag{5.24}$$

where M_b is the total mass of the beam. Equation 5.24 shows that for a beam of uniform mass distribution along its length, the modal mass equals one-third of the real mass of the beam. Note that Eqs. 5.22, 5.23, and 5.24 are only valid for linear mode shape, that is, a straight beam vibrating about its base.

Along-Wind Response of Structures due to Buffeting

The along-wind response of buildings and other structures due to buffeting by atmospheric-boundary-layer turbulence has been studied by many investigators such as Davenport (1967), Vellozzi and Cohen (1968), and Simiu (1973, 1980). Based on these studies, Solari (1982) developed a set of closed-form solutions to estimate the first mode of vibration of both rectangular structures (buildings) and small elevated blocks (pointlike structures). Due to the limited application of the latter, only the solution for rectangular structures will be discussed.

Rectangular structures (buildings). For a rectangular structure or building having height H, width (breadth) B, and depth D, with D being in the wind direction, Solari derived a set of equations and a procedure to evaluate the along-wind response due to buffeting by turbulence. His procedure is streamlined here to facilitate understanding and use by the reader. Unless otherwise indicated, all the calculations will be in SI units. The procedure now follows:

1. Determine the quantity Δ which represents the smallest of the three dimensions: B (breadth), D (depth), and H (height).
2. Determine the mass per unit length of the building $m(z)$. Also determine the air density ρ.
3. Determine the modal mass of the building, M_1, from Eq. 5.23 or 5.24.
4. Determine the fundamental (first-mode) natural frequency of the building n_1 and the damping ratio ζ.
5. Determine the windward pressure coefficient C_1 and the leeward suction coefficient C_2 which is the absolute value of the leeward pressure coefficient. Then, calculate the drag coefficient of the building from $C_D = C_1 + C_2$.
6. Determine the ground roughness z_o in meters from sources such as Table 3.1.
7. Determine the fastest 10-minute wind speed V_{10}. If the wind speed given is for a shorter averaging time, use Figure 3.4 to convert to a 10-minute average.
8. Use V_{10} in Eq. 3.1 to calculate V_*. Then calculate $q_* \rho V_*^2/2$.
9. Use the equations in Table 5.1 to calculate the intermediate quantities $F_1, F_2, \ldots, F_{10}$, and the final results $\overline{X}$ (mean vibration at the building top), X_{rms} (root-mean-square vibration at the building top), R (dominant response frequency), g (peak factor), G (gust factor), $\hat{X}$ (peak vibration), $\ddot{X}_{rms}$ (root-mean-square acceleration), g_a (peak factor for acceleration), and $\hat{\ddot{X}}$ (peak acceleration).

Note that in the foregoing procedure, V_* and q_* are calculated from the 10-minute mean velocity V_{10} which is commonly used in European practice. Instead of the 10-minute mean velocity, any mean wind speed having an averaging time T between 10 minutes and a hour may be used. The use of different averaging time needs only be reflected in the T value for calculating g_a in Table 5.1. Because the fastest-mile design wind speed, V_F, is normally much shorter than 10 minutes, the proce-

TABLE 5.1 Along-Wind Vibration of Buildings (Solari's Method)

$F_1 = 2\left(\ln \dfrac{H}{z_o}\right) - 1$	$\overline{X} = \dfrac{C_D B H q \cdot F_3}{M_1 (2\pi n_1)^2}$
$F_2 = \dfrac{n_1 H}{V \cdot F_1}$	$X_{rms} = \overline{X}\left(\dfrac{\sqrt{F_4 + F_{10}}}{F_3}\right)$
$F_3 = 0.78\, F_1^2$	$R = n_1 \sqrt{\dfrac{F_{10}}{F_4 + F_{10}}}$
$F_4 = \dfrac{6.71\, F_1^2}{1 + 0.26\,(B/H)}$	$g = \sqrt{1.175 + 2\ln(RT)}$
$F_5 = 12.32\, \dfrac{F_2 \Delta}{H}$	$G = 1 + g\,\dfrac{X_{rms}}{\overline{X}}$
$F_6 = 3.55\, F_2$	$\hat{X} = G\overline{X}$
$F_7 = \dfrac{1}{F_5} - \dfrac{1}{2F_5^2}[1 - \exp(-2F_5)]$	$\ddot{X}_{rms} = \dfrac{C_D B H q \cdot}{M_1}\sqrt{F_{10}}$
$F_8 = \dfrac{1}{F_6} - \dfrac{1}{2F_6^2}[1 - \exp(-2F_6)]$	$g_a = \sqrt{1.175 + 2\ln(n_1 T)}$
$F_9 = C_1^2 + 2C_1 C_2 F_7 + C_2^2$	$\hat{\ddot{X}} = g_a\, \ddot{X}_{rms}$
$F_{10} = \dfrac{0.59\, F_1^2 F_2^{-2/3} F_8\, F_9}{\zeta C_D^2\, [1 + 3.95\,(F_2 B/H)]}$	

dure cannot be used with V_F without first converting V_F to the mean speed of a longer averaging time, such as the hourly or 10-minute average.

The same example given by Solari (1982) will now be used to illustrate the foregoing procedure.

Example 5-3. Assume that a rectangular building has dimensions $B = 60$ m, $D = 30$ m, and $H = 200$ m. The mass per unit length of the building is $m(z) = 360,000$ kg/m. The fundamental frequency is $n_1 = 0.27$ Hz, and the damping ratio is $\zeta = 0.015$. The mean pressure coefficient on windward wall is $C_1 = 0.8$, and the mean suction coefficient on the leeward wall is $C_2 = 0.5$. The building is located in a terrain with $z_o = 7$ cm. Suppose the fastest-mile wind speed used in the design is $V_F = 72$ mph (32.2 m/s), and the air density is $\rho = 1.25$ kg/m^3. Find the mean, rms, and peak values of the along-wind vibration of the building due to buffeting.

[Solution]. Following the steps outlined before, the smallest dimension of the building is $\Delta = 30$ m, and the modal mass of the building is $M_1 = 24{,}000{,}000$ kg. For this example, the natural frequency n_1 and the damping ratio ζ are given to be 0.27 Hz and 0.015, respectively. If they are not known, they can be found in a manner to be discussed later.

Continuing the procedure yields $C_D = 1.3$, $z_o = 0.07$ m (given), $V_F = 72$ mph (given), T (from Eq. 3.11) = 50 sec, V_F/V_H (from Figure 3.4) = 1.25, $V_H = 25.8$ m/s, $V_* = 2.08$ m/s, $q_* = 2.70$ N/m, $F_1 = 14.9$, $F_2 = 1.74$, $F_3 = 173$, $F_4 = 1380$, $F_5 = 3.22$, $F_6 = 6.18$, $F_7 = 0.263$, $F_8 = 0.149$, $F_9 = 1.10$, and $F_{10} = 191$.

The foregoing procedure yields the final results $\overline{X} = 0.106$ m, $X_{rms} = 0.0243$ m, $R = 0.0941$ Hz, $g = 3.58$ (using $T = 3600$ sec), $G = 1.82$, $\hat{X} = 0.193$ m, $\ddot{X}_{rms} = 0.0243$ m/s^2, $g_a = 3.86$, and $\hat{\ddot{X}} = 0.095$ m/s^2.

Note that most of the quantities calculated in the foregoing example depend on the averaging time used. For instance, if the fastest-mile speed is converted to a 10-minute mean velocity V_{10}, instead of the hourly mean V_H, and the same procedure followed, it will be found that the mean and rms values such as $\overline{X}$, X_{rms}, and $\ddot{X}_{rms}$, will be greater by 10 to 20%, whereas the values of g, G, and g_a will be smaller by 5 to 20%. In contrast, the peak values, such as $\hat{X}$ and $\hat{\ddot{X}}$, which should not depend on the averaging time, will remain approximately constant (within 5%).

Other cantilever-beam-like structures. The along-wind response of cantilever-beam-like structures other than rectangular buildings can be predicted by the same methods for rectangular buildings. For instance, the Solari method discussed before can be used for circular cylinders such as chimneys and stacks by assuming that the leeward suction coefficient C_2 is zero, and the windward pressure coefficient C_1 is the same as the drag coefficient C_D which can be determined from the Reynolds number.

The Solari's method discussed before is unique in that it is a close-form solution which can be used easily on a computer or a programable calculator. Other methods such as those in the ANSI Standard (ANSI/ASCE 1988) and the Canadian Standard (1985) must rely on graphs and tables as well as equations. They are more difficult to program on computers or calculators.

Natural Frequency and Damping Ratio of Structures

In Examples 5-3 and 5-4, the values of the natural frequency n_1 and the damping ratio ζ were both given. In practice, n_1 and ζ must be determined for each individual building or structure.

The natural frequency of a rigidly supported cantilever beam of constant cross-sectional area and uniform distribution of material along the beam is (Hurty and Rubinstein 1964),

$$n_i = \frac{c_i^2}{2\pi H^2} \sqrt{\frac{EI}{m}} \qquad (5.25)$$

where n_i is the natural frequency of the ith mode of vibration, H is the beam length, E is the Young's modulus, I is the moment of inertia of the cross section, m is the mass per unit length, and c_i is a coefficient different for each mode. The values of c_i for the first three modes are 1.875, 4.694, and 7.855.

For the first (fundamental) mode, Eq. 5.25 reduces to

$$n_1 = \frac{0.560}{H^2} \sqrt{\frac{EI}{m}} \qquad (5.26)$$

For a rectangular structure such as a tall building, two values of n_1 exist: one for vibration along the major axis (long sides) of the rectangle, the other along the minor axis (short sides). Since the moment of inertia I is larger along the major axis, Eq. 5.26 shows that the natural frequency of vibration for a rectangular structure is higher for vibration along the major axis than along the minor axis. Because it is the lower frequency that causes greater vibration, a safe design requires consideration of the vibration along the minor axis of the rectangle.

Due to the nonuniform use of material in buildings, it is difficult to use Eqs. 5.25 and 5.26 for buildings. However, for structures such as nontapered chimneys and stacks, they can be used easily. If the chimney (stack) diameter and wall thickness are D and t, respectively, the moment of inertia can be calculated from

$$I = \frac{\pi t D^3}{8} \qquad (5.27)$$

Equations 5.25–5.27 provide a means to determine the natural frequencies of chimneys and stacks.

The natural frequencies of buildings are normally determined from empirical relations. Ellis (1980) analyzed the test results for a large number of buildings and found that n_1 can be determined approximately from

$$n_1 = \frac{46}{H} \qquad (5.28)$$

where n_1 is in hertz and H is in meters. He found that this simple formula, though it may yield errors in excess of 50%, is as accurate as or more accurate than contemporary, computer-based methods. It should be mentioned that Eq. 5.28 is somewhat at odds with the commonly accepted rule-of-thumb formulas of $n_1 = 100/H$, where H is in feet, and $n_1 = 10/N$, where N is the number of stories above ground. It is believed that Eq. 5.28 is more accurate than these formulas because it was derived statistically from a large number of field data.

The magnitude of the damping ratio ζ depends not only on the structure itself but also on the structure-soil interaction. Even for a given structure with a given foundation, the value of ζ still changes with changing magnitude of vibration. At larger amplitudes of vibration, aeroelastic damping can become significant, and the total damping ζ can be considered as the sum of the structural damping ζ_s and the aeroelastic damping ζ_a. The current state of the art in predicting ζ is rather crude; it is not uncommon to have predictions in error by more than 100%. The normal range of ζ for structures vibrating at low amplitudes is 0.005–0.025. Without better information, it is common for designers to use $\zeta = 0.01$ for steel frame structures and $\zeta = 0.02$ for concrete structures. More about damping for tall buildings can be found in many publications such as Jeary and Ellis (1983).

Chimneys and stacks have lower damping than buildings. Typical ranges of ζ are 0.002–0.008 for unlined steel stacks, 0.003–0.010 for steel stacks with 2-in. (5cm) lining, 0.005–0.012 for steel stacks with 4-in. (10 cm) lining, and 0.004–0.020 for reinforced concrete chimneys. Lower values of each range should be used for chimneys and stacks supported on rock foundation or low-stressed piles, medium values for soil foundation, and higher values for roof-supported stacks and chimneys of thick walls.

Across-Wind Response of Structures

Vortex shedding often induces strong vibration of structures in the across-wind direction. The severity of such self-induced vibration depends on whether lock-in plays a role or not. When lock-in is present, vibration is essentially sinusoidal and its amplitude is very large. On the other hand, when lock-in is absent, vibration is random and its amplitude is relatively small.

Whether lock-in is present or not in a given case depends on whether the intensity of across-wind vibration at the structure top, Y_{rms}, has reached a critical value Y_c. The value of Y_c depends on building

Sec. 5.4 Predicting Dynamic Response of Structures

geometry and the turbulence in the wind. Wind tunnel test results have indicated that for buildings of square cross section having width B, the values of Y_c/B are approximately 0.015, 0.025, and 0.045, respectively for open terrain ($z_o = 0.07$ m), suburban terrain ($z_o = 1.0$ m), and city centers ($z_o = 2.5$ m). For circular cylinders of diameter D, the value of Y_c/D is approximately 0.006 for suburban terrain ($z_o = 1.0$).

Across-wind vibration of structures is caused by the combined forces from three sources: (1) buffeting by turbulence in the across-wind direction, (2) vortex shedding, and (3) aeroelastic phenomena such as lock-in or galloping. Due to the complex interaction among these forces, current knowledge in this field is incomplete. All current methods for predicting across-wind vibration rely heavily on wind tunnel data. The following is a method that can be used for buildings.

Australian standard method. The old Australian Standard (1983 and before) includes a simple formula proposed by Vickery to estimate the across-wind response of buildings produced by winds. The formula, written here in a slightly different form, is

$$Y_{\text{rms}} = 0.00015 \frac{\rho}{\rho_b} \sqrt{\frac{A}{\zeta_1}} \left(\frac{V_H}{n_1 \sqrt{A}}\right)^{3.5} \qquad (5.29)$$

where ρ is the density of the air, ρ_b is the bulk density of the building, A is the cross-section area of building, ζ is the damping ratio, V_H is the hourly mean wind speed at roof height, and n_1 is the fundamental natural frequency of building vibration in the across-wind direction.

Note that Eq. 5.29 can be used for a large variety of cross-sectional shapes of buildings such as circle, rectangle, square, and triangle. The Y_{rms} calculated is that produced by winds from the most adverse direction. For instance, for square cross sections, the most adverse direction of wind for across-wind vibration is when the wind is perpendicular to a wall. For a rectangular building, the most adverse is when the wind is perpendicular to the short side of the building (i.e., depth D is greater than breadth B).

Once the root-mean-square of Y has been calculated from Eq. 5.29, the peak value $\hat{Y}$ can be taken as four times the rms value. The root-mean-square of $\ddot{Y}$ (acceleration) can be calculated from

$$\ddot{Y}_{\text{rms}} = (2\pi n_1)^2 \, Y_{\text{rms}} \qquad (5.30)$$

The peak acceleration can be taken as four times the rms value.

The 1989 edition of the Australian Standard uses a different equation than Eq. 5.29. This new method relies on a cross-wind spectrum coefficient that must be determined from a set of graphs for different building dimensions $H: B: D$.

Example 5-4. Determine the across-wind response of the building described in the previous example.

[Solution]. From the previous example, $\rho = 1.25$ kg/m^3, $\rho_b = 200$ kg/m^3, $A = 60 \times 30 = 1800$ m^2, $\zeta = 0.015$, $V_H = 25.8$ m/s, and $n_1 = 0.27$ Hz. Substituting these values into Eq. 5.29 yields $Y_{rms} = 0.00557$ m and $\hat{Y} = 4Y_{rms} = 0.0223$ m. Then, from Eq. 5.30, $\ddot{Y}_{rms} = 0.0160$ m/s^2 and $\hat{\ddot{Y}} = 4\ddot{Y}_{rms} = 0.0641$ m/s^2.

Comparing the values of the across-wind response found in this example with the corresponding values of the along-wind response found in the previous example shows that the along-wind response is more serious than the across-wind response for this building.

Note that for the along-wind response calculations in Example 5-3, the wind is perpendicular to the long sides (60-m sides) of the building. In contrast, the across-wind response analysis in Example 5-4 assumes that the wind is perpendicular to the short sides (30-m sides). To combine the across-wind effect with the along-wind effect, one must use the wind from the same direction in both analyses. If we assume the wind to be perpendicular to the long sides of the building, the natural frequency n_1 used for analyzing the across-wind vibration should be higher than 0.27, which is the vibration frequency along the short sides (minor axis).

Vibration of Stacks, Chimneys, and Towers

Steel stacks, concrete chimneys, and towers of circular cross section are most susceptible to self-induced (vortex-shedding generated) across-wind vibration. A simple method for estimating the response of such structures is described next.

Canadian method. Commentary B to the Supplement of the National Building Code of Canada (NBC Commentary 1985) gives the following simple formula for predicting the equivalent static load for an untapered stack, chimney, or circular tower,

$$w = \frac{Cq_H D}{\sqrt{\lambda[(\zeta - (0.6\rho D^2/m)]}} \quad (5.31)$$

where w is the equivalent static load per unit length of the structure, to

Sec. 5.4   Predicting Dynamic Response of Structures   139

be applied only to the top one-third of the stack; D is the diameter of the structure; λ is the aspect ratio H/D where H is the structure height; q_H is the velocity pressure (dynamic pressure) based on the mean hourly wind speed at stack height; ζ is the damping ratio; ρ is the air density; m is the mass per unit length over the top one-third of the structure; and C is equal to 3.0 when the aspect ratio λ is greater than 16 and equal to 0.75 $\sqrt{\lambda}$ for λ less than 16.

Equation 5.31 holds only for ζ greater than $0.6\rho D^2/m$. Otherwise, large amplitude vibration up to one structure diameter may occur, requiring a redesign of the structure (vertical cylinder).

If the vertical cylinder is tapered, but the variation in diameter over the top third is less than 10% of the average diameter, the same load as for the untapered structure is used, except that D in Eq. 5.31 must now represent the average diameter. If the diameter variation exceeds 10%, then the equivalent static load is to be applied only over that part of the structure where the diameter is within 10% of the average for that part. For a tapered structure with a diameter variation exceeding 10% over the top third, $C = 3$ is used for all values of λ.

Example 5-5.   An untapered reinforced-concrete chimney has a diameter of 4 m, a height of 100 m, and a thickness of 15 cm. Determine the maximum deflection of its top under a 20-m/s hourly wind.

[Solution].   For this chimney, $D = 4$ m, $H = 100$ m, $t = 0.15$ m, and $V_H = 20$ m/s. The aspect ratio $\lambda = H/D = 100/4 = 25$. Since λ is greater than 16, we have $c = 3.0$. Furthermore, the density of the concrete is $\rho_c = 2400$ kg/m^3, the mass per unit length of the chimney is $m = \rho_c \pi D t = 4524$ kg/m, the density of air is $\rho = 1.2$ kg/m^3, the dynamic pressure is $q_H = \rho V_H^2/2 = 240$ Pa, and the damping ratio is $\zeta = 0.015$ (approx.).

Substituting the foregoing values into Eq. 5.31 yields $w = 5162$ N/m. Since this load is to be distributed over the top one-third of the chimney, the action is approximately the same as that of a load $F = wH/3 = 172{,}000$ N, concentrated at a height $b = (5/6)H = 83.3$ m from the ground. The deflection of the top of the beam (chimney in this case) due to this equivalent static load can be calculated from the following formula:

$$\delta = \frac{Fb^2\,(3H - b)}{6EI} \tag{a}$$

From Eq. 5.7, the moment of inertia of the chimney cross section is

$I = 3.77 \text{ m}^4$. Assume the concrete has a Young's modulus of $E = 2.2 \times 10^{10}$ Pa. From Eq. (a), $\delta = 0.52$ m. Therefore, the peak deflection caused by across-wind vibration of this chimney is $\hat{Y} = \delta = 0.52$ m.

5.5 WIND-INDUCED FATIGUE

Introduction

Fatigue is the fracture of structural components caused by a number of load fluctuations or reversals. The repeated stress reversals in materials causes minor cracks to accumulate and to develop into large cracks, which weaken the materials. Generally, a much smaller stress is required to cause material fracture by fatigue than by a static load. Thus, fatigue failure plays a major role in wind damage to buildings and structures—more than normally realized.

It is a common practice to classify fatigue either as high-cycle fatigue or low-cycle fatigue. High-cycle fatigue is defined as the fatigue caused by more than 10^4 cycles of load, whereas low-cycle fatigue occurs within 10^4 cycles. Stresses that cause low-cycle fatigue are normally high enough to cause considerable plastic deformation. In such a situation, it is usual to consider the cyclic strain rather than the stress as a function of fatigue cycles.

Fatigue is caused by the development of innumerable small cracks in material, growth of these cracks with each additional cycle of stress, and the combination of small cracks to form large cracks that lead to ultimate failure. The damage is cumulative. To predict the fatigue life of a specimen under a random or irregular load, the linear cumulative theory (the Palmgrem-Minor theory) can be applied. The theory says that fatigue is proportional to the accumulated damage done to the specimen. Therefore, the damage accumulated from any level of load is equal to the ratio of the number of load cycles n that the specimen has been subjected to and the number of cycles N required to produce failure. Mathematically, fatigue failure occurs when

$$\sum_{i=1}^{k} \left(\frac{n}{N}\right)_i = 1 \qquad (5.32)$$

where each i represents a different load level. It should be realized that Eq. 5.32 is not exact; it yields only the correct order of magnitude of the fatigue life.

The fatigue property of structural components can be found

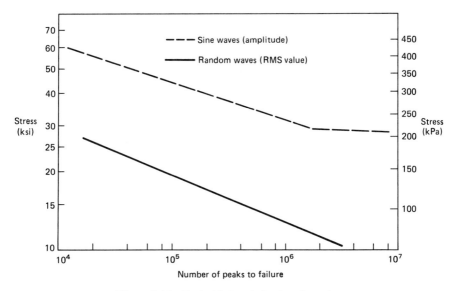

Figure 5.11 Typical fatigue behavior of metals.

through tests. These tests may be conducted either with constant-amplitude load (sinusoidal load) or variable-amplitude load. Variable-amplitude load tests can be further divided into programme-load test and random-load test. A programme-load test involves repeating a block of specified loads, each containing a given number and sequence of cycles of loads of various amplitudes. To simulate a random load, each block should have of the order of one thousand cycles. The programme-load test is more common than the random-load test; it is routinely used in the aerospace industry. It is easier to generate a programme load than a random load. The load used in fatigue tests can be provided by servovalve-controlled hydraulic equipment.

Figure 5.11 illustrates a typical fatigue test result. It shows that the greater the number of cycles of loading a material has been subjected to, the smaller a stress is required to fracture the material. Usually, random loads are more critical than sinusoidal (sine-wave) loads in causing fatigue.

Wind-Induced Fatigue

The subject of wind-induced fatigue on structures has not been thoroughly explored to date. Only limited knowledge exists in the field. It is anticipated that both low-cycle fatigue and high-cycle fatigue can be produced by high winds. In the case of tornado and thunderstorm straight-line winds, the duration of wind is short and the stress generated is high. Thus, low-cycle fatigue may dominate in this case. In areas frequented by tropical cyclones (hurricanes, typhoons, etc.) or by mountain downslope winds, both low- and high-cycle fatigues may play a role. A simple example is now considered:

Suppose a given structure or structural component has a natural frequency of 30 Hz. Suppose the structure is exposed to a strong hurricane wind for 10 hours. During this time, the structure may undergo as many as $30 \times 3600 \times 10 \simeq 1,080,000 = 10^6$ cycles of vibration. Since this falls within the range of high-cycle fatigue, even moderate stresses produced by vibration can cause the structure or component to fail within the 10 hours of the hurricane. This points to the seriousness of allowing structures to vibrate in a hurricane wind, even when the stress produced by vibration is much less than the allowable stress used in design. The structure is even in greater peril if it is situated in an area frequented by hurricanes or mountain downslope winds. Lynn and Stathopoulos (1985) presented an analytical approach for the evaluation of wind-induced fatigue on low-rise metal buildings.

Case Study

Australian experience. The northern coast of Australia was struck by a cyclone named Tracy on Christmas day of 1974, the city of Darwin suffering great damage. According to Walker (1975), between 50 and 60% of timber houses were damaged beyond repair. Over 90% of houses and 70% of other structures suffered a significant loss of roofing. Fatigue failure of roof joints was believed to be a major cause of damage. This was especially serious for corrugated roof—a common roofing material used in the region.

As a result of the lesson learned from cyclone Tracy, the Darwin Area Building Manual incorporated a provision that required fatigue tests of roofs. Prototypes of roof assemblies must be subjected to 10,000 cycles of repeated load, with the load varying from zero to the design load corresponding to a 50-year return period. Following the repeated load test, the roof assembly was to be subject to a static load equal to 1.8 times the design load. Later studies showed that the requirements were overly conservative. The 1989 edition of the Australian standard specifies the fatigue loads that claddings and connections must be designed to resist in cyclone regions.

Research was conducted at the University of Melbourne (Beck

Sec. 5.5 Wind-Induced Fatigue

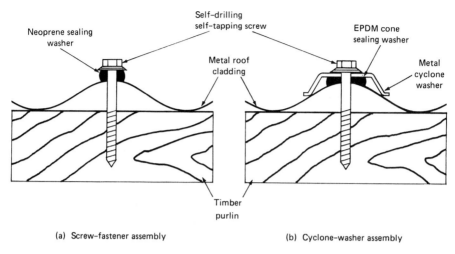

Figure 5.12 Comparison of cyclone washer with ordinary washer for corrugated roof. (From Beck, 1978.)

1978) in which a corrugated steel roof was tested for its fatigue properties. The roof assembly was constructed in such a way to duplicate the typical corrugated roof used in Darwin. A set of programme-load tests were carried out. An important finding was that the fatigue life of the roof depended greatly on the screws which fastened the corrugated steel to the purlins. The tighter the screws were fastened, the longer the life of the roof. It was found that by using a cyclone washer in association with each screw (see Figure 5.12), the fatigue life of the roof assembly can be extended 100 times! This study demonstrated the great importance of using proper connectors (especially washers) to attach roofing materials, and the importance of tightening connectors.

Kemper Arena. The Kemper Arena is a large auditorium in Kansas City, Missouri, that houses sporting and civic events. On June 4, 1979, a local thunderstorm caused a large portion of the roof to collapse. Fortunately, no events were scheduled and no one was under the roof at the time. In spite of that, the damage exceeded $5 million and initiated a number of lawsuits.

A postdisaster investigation (Stratta 1980) concluded that fatigue failure of one of the large bolts that connected the roof to the space frame above the roof caused progressive failure of the roof. It is one of the few cases in which fatigue has been identified as the cause of failure of a major building.

Tacoma Narrows Bridge. The Tacoma Narrows Bridge in the state of Washington collapsed in 1940 only four months after the bridge had been opened. Large-amplitude vibration of the bridge, both in linear and torsional modes, had been observed prior to the collapse. Although most studies of this failure case centered on vortex shedding, bridge stiffness, and the types of vibration, it is another classical example of fatigue-caused failure.

In conclusion, it may be said that wind-induced vibration of any kind, if allowed to continue over a long period, will inevitably lead to fatigue failure of the structure. Larger vibration will cause fatigue failure to occur earlier. Structural connections (joints) are especially vulnerable to fatigue failure. Undoubtedly, increased emphasis on fatigue considerations will be required in structural design in the future.

Mitigation of Wind-Induced Fatigue

Since fatigue is produced by vibration, the most effective way to mitigate fatigue damage is to eliminate or reduce wind-induced vibration. All the measures discussed in Section 5.3, such as using helical strakes to protect stacks, using fairing to increase the aerodynamic stability of certain bridge sections, and using tuned-mass or viscoelastic dampers to reduce building vibration, are applicable to combating fatigue problems. For connectors such as bolts and screws, using large washers to distribute loads to a larger area and tightening the connectors are all effective in controlling fatigue failure. Using fatigue resistant materials may also provide a possible solution. Since fatigue takes time to develop, periodic inspections of structures that have been subjected to wind-induced vibration to detect fatigue cracks are important to safeguard structural integrity.

REFERENCES

ANSI A58.1 (1982). *Minimum Design Loads for Buildings and Other Structures*, American National Standard Institute, New York.

ANSI/ASCE 7 (1988). *Minimum Design Loads for Buildings and Other Structures*, American Society of Civil Engineers, New York.

Australian Standard (1983). *SAA Loading Code*. Part 2: *Wind Forces*, Standard Association of Australia, North Sydney.

Australian Standard (1989). *SAA Loading Code*. Part 2: *Wind Loads*, Standard Association of Australia, North Sydney.

BECK, V. R. (1978). *Wind Load Failure of Corrugated Roof Cladding*, M. S. Thesis, Department of Civil Engineering, University of Melbourne, Victoria, Australia.

BLEVINS, R. D. (1977). *Flow-Induced Vibration*, Van Nostrand Reinhold Company, New York.

CANADIAN STANDARD (1985). *National Building Code of Canada*, National Research Council of Canada, Ottawa.

DAVENPORT, A. G. (1967). "Gust Loading Factors," *Journal of the Structural Division*, ASCE, 93(3), 11–34.

ELLIS, B. R. (1980). "An Assessment of the Accuracy of Predicting the Fundamental Natural Frequencies of Buildings and the Implications Concerning the Dynamic Analysis of Structures," *Proceedings, Inst. Civil Engrs.*, Part 2, 69, 763–776.

HURTY, W. C. AND RUBENSTEIN, M. F. (1964). *Dynamics of Structures*, Prentice-Hall, Englewood Cliffs, New Jersey.

JEARY, A. P. AND ELLIS, B. R. (1983). "On Predicting the Response of Tall Buildings to Wind Excitation," *Journal of Wind Engineering and Industrial Aerodynamics*, 13, 173–182.

LYNN, B. A. AND STATHOPOULOS, T. (1985). "Wind-Induced Fatigue on Low Metal Buildings," *Journal of Structural Engineering*, ASCE, 111(4), 826–839.

NBC Supplement B (1985). *Supplement to National Building Code of Canada, Commentary B, Wind Loads*, National Research Council of Canada, Ottawa.

SCANLAN, R. H. AND WARDLAW, R. L. (1973). "Reduction of Flow-Induced Structural Vibrations," *Isolation of Mechanical Vibration, Impact, and Noise*, AMD Vol. 1, Sec. 2, ASME, New York.

SIMIU, E. (1973). "Gust Factors and Alongwind Pressure Correlations," *Journal of the Structural Division*, ASCE, 99(4), 773–782.

SIMIU, E. (1980). "Revised Procedure for Estimating Along-Wind Response," *Journal of the Structural Division*, ASCE, 106(1), 1–10.

SIMIU, E. AND SCALAN, R. H. (1986). *Wind Effects on Structures* (2nd ed.), John Wiley, New York.

SOLARI, G. (1982). "Along-Wind Response Estimation: Closed-Form Solution," *Journal of the Structural Division*, ASCE, 108(1), 225–244.

STRATTA, J. L. (1980). *Report of the Kemper Arena Roof Collapse of June 4, 1979, Kansas City, Missouri*, James L. Stratta Consulting Engineer, Menlo Park, California.

VELLOZZI, J. AND COHEN, E. (1968). "Gust Response Factors," *Journal of the Structural Division*, ASCE, 94(6), 1295–1313.

WALKER, G. R. (1975). *Report on Cyclone Tracy Effect on Buildings*, Vol. 1, Australian Department of Housing and Construction, Melbourne.

6

Wind Tunnel Tests

6.1 INTRODUCTION

As can be seen from earlier chapters, there are many situations in which the design wind load or the response of flexible structures cannot be predicted with sufficient accuracy either to assure safety or to avoid using uneconomically large safety factors (ultraconservative designs). In such situations as well as in many other circumstances that will be discussed in this chapter, it may be desirable to conduct wind tunnel tests of structural models. The purpose of wind tunnel tests is to provide designers with information on local wind patterns, wind loads, and wind-induced structural vibration having an accuracy far exceeding that can be obtained from predictions based on other less expensive means such as theory, numerical analysis, expert judgment (consulting), and so on.

Structures that often warrant a wind tunnel test include skyscrapers, large domes, long-span bridges, large structures of unusual shapes or of unusual flexibility and light weight, major structures in special locations affected by topographical features such as hills, cliffs, valleys, canyons, and so on. In all these as well as a number of other situations, the use of wind tunnel tests to improve design is either desirable or necessary. Failure to conduct a wind tunnel test in such a situation may result in either an unsafe design which increases the liability of the designer or an overly conservative design costly to the owner.

Due to the foregoing, all structural engineers and architects need to have some knowledge in wind tunnel testing of structures, so that they can decide whether in a given situation a wind tunnel test should be commissioned. They also must know what types of wind tunnel tests are needed, and must be able to communicate effectively with the wind tunnel test providers so that any test conducted will be of maximum value to the designer.

6.2 CIRCUMSTANCES FOR CONDUCTING WIND TUNNEL TESTS

The use of wind tunnels to aid in structural design and planning has been steadily increasing in recent years. A question often asked by structural engineers, architects, and city planners concerns when a wind-tunnel model test should be conducted to facilitate design or planning. This question defies a simple answer, for it depends on a number of factors such as the cost of the structures, the anticipated likelihood that the structure may develop wind problems after it is built, the complexity of the structure (i.e., whether the wind load or the response of the structure can be predicted accurately without a wind tunnel test), the importance of the structure (how critical it is if the structure is failed by winds or if it develops wind problems), and finally, the performance criteria of the structure as specified by codes and standards or by the structure's owner or financier. All these factors must be considered to determine whether a wind tunnel test is desirable or necessary. An elaboration of these factors is given next.

Cost of Structures

Depending on the complexity of the study, a wind tunnel model test conducted in the United States normally costs at least a few thousand dollars, more often in the range of $10,000 to $50,000 (1990 price). Obviously, one cannot afford or justify a wind tunnel test for every building. Since the most common incentive for wind tunnel tests is cost saving, a wind tunnel test cannot be justified unless the expected saving from such a test is greater than the cost for conducting the test.

Safety is usually not the motivation for wind tunnel tests because when safety is uncertain in a given case, the designer can always use stronger structural members, stronger connections, or better materials, instead of a wind tunnel test. Only if a wind tunnel test is less expensive than the other means to ensure the same degree of safety will the test be commissioned by the designer. Because the potential for cost saving from wind tunnel model tests is great for expensive structures, it is the multi-million-dollar structures that usually require such tests.

Likelihood of Wind Problems

When a designer feels that there is a good chance that a structure may run into wind problems without a better knowledge of the wind field, pressure distribution, or the structural response to wind, a wind tunnel model test should be considered. Conditions conducive to wind problems include special geographical locations (e.g., a skyscraper in a hurricane-prone area or a large dome in a region affected by mountain downslope winds), special topography (e.g., a building on the top of a hill), and wind-sensitive structures (e.g., tall buildings and long-span bridges).

Complex Structures

Structures of unusual shape may require a wind tunnel model test, because the distribution of wind pressure on such structures and the flow pattern around them may not be given by building codes and standards and may not be known from any other source. In such a case, a wind tunnel test is the only way to generate the information on wind load and wind pattern needed for design.

Importance of Structures

A building that houses thousands of people obviously requires greater protection than a warehouse that stores lumber. Even though building codes and standards already require a higher wind load for more important structures, the designer and the owner may like to exceed the design loads required by codes/standards for particularly important buildings in order to minimize the risk of large potential life loss and the associated liability. Using wind tunnel data in design often contributes to a safer design, without having to make the building unnecessarily costly.

Performance Criteria for Structures

Some special structures may have to satisfy special performance criteria either dictated by the function of the structures, by codes or standards, or by the owner's wish. For instance, while cladding failure such as wind damage to a large glass window of an ordinary building may not be a serious problem, the same does not hold for a structure that houses important sensitive equipment that must be protected from damage or disruption by wind storms, as in the case of an aviation traffic control tower, or the computer center of a telephone company. Such important buildings have special performance criteria that must be met by the designer. A wind tunnel model test may be needed in such a case to determine the load accurately so that the building can be designed to satisfy such special performance criteria. Some building performance criteria are set by the owner. For instance, the owner of a tall office building may like to set a serviceability criterion such as the vibration of the building must not exceed a certain amount more than once a year. A wind tunnel test of an aeroelastic model of the building will enable the designer to see whether his or her initial design satisfies the owner's criterion.

Some of the types of most commonly conducted wind tunnel model tests are the following:

1. Test of a tall building surrounded by other tall buildings in an urban setting—the interference effect of tall buildings. In this case, both a rigid model and an elastic model are usually tested. The rigid model is for determining the distribution of wind pressure (external pressure) on the cladding of the proposed building and the wind field—speed, direction, and flow pattern around the building, especially at the pedestrian level. The elastic model is for building vibration study to determine dynamic wind load and serviceability condition.

2. Test of a large dome or a building having a large roof with a long span. The main purpose is to assess the wind load on the roof accurately, for the roof load governs the design in this case. Flow patterns around such buildings is also often of interest for determining wind effects on people entering and leaving the building, on opening and closing doors on windy days, and on operation of ventilation, cooling, and heating systems of such buildings.

3. Test of bridge models (sectional models, full-span models, or com-

plete bridge models) to determine wind-induced vibration of cable-suspended, long-span bridges.
4. Test of special structures such as an off-shore oil platform, a tall monument, a radar station, the launch facility of the space shuttles, and so on, for assessing potential wind problems.
5. Tests to determine the effect of roof-level windiness on the operation of helicopters on roofs and, more commonly, on the adequacy of ventilation inlets and outlets on roofs, and on the proper operation of roof-mounted cooling towers and air conditioning equipment.
6. Tests to determine snow drift on roofs having unusual geometry.
7. Tests to determine the amplification or alteration of wind by hills and other geographical features.
8. Test of power plants, chemical plants, and factories, to assure proper dispersion of pollutants emitted from such plants or factories.
9. Test of urban wind environment to devise strategies to combat air pollution.

Now that the reader has some idea as to when a wind tunnel model test should be conducted, he or she should have some knowledge about the types of wind tunnel available and the key elements involved in wind tunnel modeling of structures, so that he or she can effectively communicate with the wind tunnel test provider and use the test results more effectively. For this reason, he or she should read the rest of this chapter.

6.3 TAXONOMY OF WIND TUNNELS

Numerous criteria exist to categorize wind tunnels. In what follows, various wind tunnels will be discussed under various classification headings.

Flow Circuit

According to flow circuit, a tunnel may be classified either as an **open-circuit** or a **closed-circuit type**. An open-circuit tunnel is normally a straight structure. Air is drawn into the tunnel from a funnel-shaped intake at one end of the tunnel, and the air exits the tunnel through a funnel-shaped outlet. Figure 6.1 is a sketch of such a tunnel.

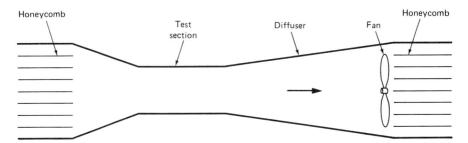

Figure 6.1 Open-circuit wind tunnel.

The enlarged cross-sectional areas at the two ends reduce head loss (energy dissipation) and prevent undesirable strong winds from being generated outside the tunnel near the inlet and outlet.

The closed-circuit type is a recirculating loop as shown in Figure 6.2. Air is circulated through the loop during tests. A closed-circuit tunnel may occupy a large space if the loop is horizontal. Consequently, indoor closed-circuit tunnels are sometimes arranged in a vertical loop to save laboratory space. Some large tunnels of the closed-circuit type utilize the enlarged return section as an additional test section for low-speed tests such as those required for air pollution or wind energy studies.

Advantages of closed-circuit tunnels are (1) they do not cause undesirable wind in laboratories housing the wind tunnels; (2) they

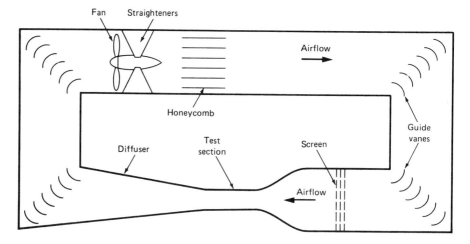

Figure 6.2 Closed-circuit wind tunnel.

Sec. 6.3 Taxonomy of Wind Tunnels

Figure 6.3 Aeronautical wind tunnel at the Ames Research Center, National Aeronautic and Space Administration (NASA). (This wind tunnel, the largest in the world, has an 80-ft × 120-ft test section large enough for testing full-size small aircraft.) (Courtesy of NASA.)

generate less noise in the laboratory; and (3) when placed outdoors, they do not suck rain, snow, and dust into the tunnel. It is mainly for this last reason that most outdoor tunnels are of the closed-circuit type. An example of a large outdoor wind tunnel is NASA's Ames wind tunnel in California—the largest in the world (see Figure 6.3).

Both open- and closed-circuit tunnels are often used for testing structural models.

Throat Condition

The throat of a wind tunnel is the test section. It may either be closed or open to the environment. Both the tunnels in Figures 6.1 and 6.2 are of the closed-throat type. In contrast, Figure 6.4 shows an open-throat or open-jet tunnel in which models are tested in the open section. Certain tests such as those involving aircraft models must be tested in uniform velocity and atmospheric pressure—conditions that can be provided by an open-throat arrangement. Note that open-throat tunnels are for aeronautic and basic fluid mechanic studies; they are not

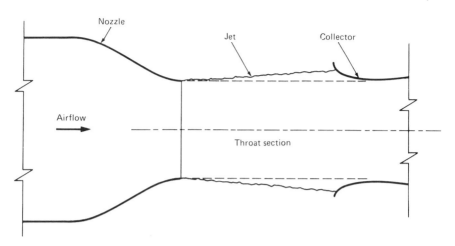

Figure 6.4 Open-throat wind tunnel.

suitable for testing civil engineering structures such as buildings and chimneys, which are exposed to turbulent boundary-layer winds. Testing of most civil engineering structures requires a boundary-layer wind which cannot be generated in an open-throat section.

Pressure Condition

Most wind tunnels are not pressurized, even though their pressure varies from place to place along the tunnel axis and differs somewhat from the atmospheric pressure outside. The pressure in a wind tunnel is the highest immediately downstream of its fan. It reaches a minimum on the suction side of the fan. Due to the speed-up of flow through the throat (test section), the pressure in the throat is normally low—less than atmospheric.

Special pressurized tunnels are those needed for generating very high Reynolds number. As will be discussed later, for certain types of structures it is important to run the model tests at the same Reynolds number as encountered by the prototype. To do that, one must either increase the wind speed used in the model tests or increase the density of air through compression. Since increasing wind speed can cause a mismatch in the Mach number, the only practical solution is compressing the air in the wind tunnel, or using a fluid of much higher density than air—liquid. To be effective in increasing Reynolds number, a tunnel must be compressed to rather high pressure, say, 50 atmospheric pressure (735 psi or 5070 kPa). Only a few such tunnels exist (or

Sec. 6.3 Taxonomy of Wind Tunnels

existed) in the world to study the Reynolds number effect. An example of such a tunnel is one at the Jet Propulsion Laboratory of the California Institute of Technology used to study flow around circular cylinders at high Reynolds numbers (Roshko 1961). Due to the high cost of large-size pressurized tunnels, tests of buildings and other structures are generally done in nonpressurized tunnels.

Wind Speed

High-speed tunnels are those having a Mach number approaching or exceeding one—the sonic velocity. The flow in such tunnels is considered "compressible," which means the density of air changes significantly from place to place in the tunnel or around test objects. In low-speed tunnels (Mach number less than one-third), the density of air is essentially constant everywhere, and the flow can be considered "incompressible."

Wind tunnels are commonly referred to as **low-speed tunnels** if the Mach number is less than one-third which under atmospheric condition corresponds to a speed of 110 m/s (250 mph) approximately. They are referred to as **high-speed tunnels** if the Mach number is greater than one-third, as **supersonic tunnels** if the Mach number is greater than 1.0, and as **hypersonic tunnels** if the Mach number is greater than 3.0.

Wind tunnels for testing structural models are low-speed tunnels of large cross-sectional area. They are normally kept within a speed of 50 m/s (112 mph), often within 25 m/s (56 mph), to minimize cost. As will be shown later, wind speeds greater than 10 m/s are normally unnecessary for testing structural models.

Velocity Profile

Tunnels for aeronautical applications and for testing automobiles must have uniform velocity and smooth flow (nonturbulent flow) at the test section. The velocity profile must be uniform because the relative velocity between a flying aircraft (or a moving automobile) and the surrounding stationary air is uniform (constant).

In contrast, for testing structural models we must generate vertical distribution of velocity at the tunnel test section similar to the logarithmic profile or the power-law profile of the wind encountered by prototypes. Also, the turbulence in the wind must be simulated correctly. This calls for a special type of wind tunnel for structural testing: the **boundary-layer wind tunnel**. Note that prior to 1955, wind tunnel tests of structural models were often conducted in aeronautical tunnels of smooth, uniform, flow. Results of such tests should not be used without great caution. Contemporary tests of structural models, except for bridge models, are almost always conducted in boundary-layer tunnels. Due to their importance to structural engineering, boundary-layer wind tunnels will be discussed in more detail under a separate heading.

Temperature Stratification

Most tunnels generate wind of constant (uniform) temperature. However, some **meteorological tunnels** have been built to study air pollution (plume dispersion) which requires vertical stratification of temperature. For instance, Colorado State University has such a tunnel; see Figure 6.5. For testing wind loads on structures, temperature stratification in wind tunnels is not needed. It is generally believed that due to intense mixing, little temperature stratification exists in high winds.

Cross-sectional Geometry

Most tunnels have rectangular or square cross sections; a few have circular cross sections. For aeronautic or other purposes that require a uniform velocity, the shape of the tunnel is unimportant. In contrast, boundary-layer tunnels must have rectangular or square cross sections to simulate properly the variation of velocity with height in the atmospheric boundary layer.

Drive System

Most tunnels are driven by fans. A few are driven by wall jets approximately parallel to the mean flow direction. Due to the low efficiency of energy conversion from jets to the mean flow, jets are not used for driving wind tunnels except in special situations.

Turbulence Level

The turbulence characteristics in a wind tunnel can be controlled by a number of means such as by installing screens, grids, and spires upstream of the test section; by placing roughness elements on the floor of tunnels; by using counter jets; and so on. For structural testing, the relative turbulence intensity I_r is normally in the range 10–30%. Some tunnels have been designed to have very low turbulence level (less than

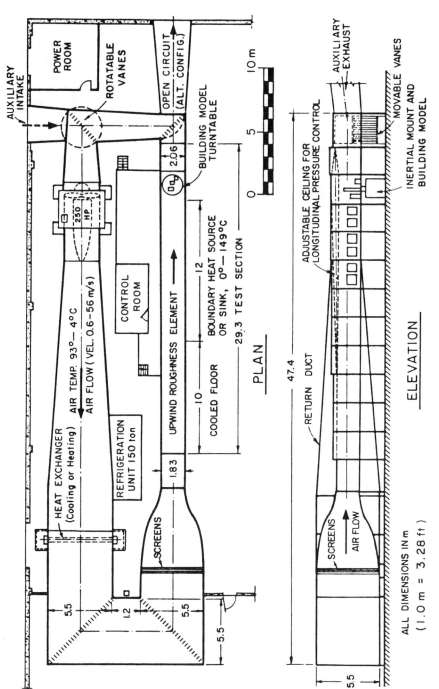

Figure 6.5 Meteorological wind tunnel at the Fluid Dynamics and Diffusion Laboratory, Colorado State University. (Courtesy of the Laboratory.)

Figure 6.6 Water-wave wind tunnel at the Boundary-Layer Wind Tunnel Laboratory, University of Western Ontario, Canada. (Courtesy of the Laboratory.)

1% or sometimes even less than 0.1%). Such tunnels are for special purposes such as studying smooth, uniform flow around circular cylinders.

Special-Purpose Tunnels

There are various types of special tunnels for special purposes. For instance, smoke tunnels are those needed for flow visualization. Such tunnels are equipped with smoke generators and exhaust fans.

Wave tunnels are those with a water flume on the floor of the wind tunnel. They are used for studying wind-generated waves and related phenomena, such as ocean waves hitting an off-shore platform; see Figure 6.6. Many experts believe that it is very difficult to get both the winds and waves simulated properly simultaneously. Most frequently, the tests are conducted separately and results superimposed.

Special tunnels were also built to study bridge models. Most **bridge tunnels** use smooth, uniform wind; they are characterized by wide test sections. Some have the capability of generating free-stream turbulence for studying the effect of turbulence on the response of bridges to wind.

Figure 6.7 Example of a commercial wind tunnel for structural testing. The bridge is in the front and the spires in the background. (Courtesy of Wind Dynamics Laboratory, Applied Research Associates, Raleigh, North Carolina.)

6.4 BOUNDARY-LAYER WIND TUNNEL

A boundary-layer wind tunnel must have a test section that is sufficiently long to generate a thick vertical boundary layer, sufficiently high so that the boundary layer generated will not touch the tunnel ceiling, and sufficiently wide so that neighboring structures and topographical features can be incorporated into the model. Furthermore, the blockage ratio (i.e., the ratio of the cross-sectional area of the model blocking the flow and the cross-sectional area of the tunnel test section) must be less than approximately one-tenth. These requirements necessitate rather large tunnels. The boundary-layer tunnels used in commercial testing of structural models normally have a minimum width of 2.4 m (8 ft), minimum height of 1.5 m (5ft), and a minimum length of 10 m (33ft)—all referring to test sections. Figures 6.7 and 6.8 are two examples.

Figure 6.8 Example of a commercial wind tunnel for structural testing. (Courtesy of Cermak/Peterka/Petersen, Inc., Fort Collins, Colorado.)

To facilitate the rapid growth of a vertical boundary layer along the tunnel test section, not only must roughness elements be placed on the tunnel floor, but additional devices such as spires (see Fig. 6.9) must be installed upstream. The roughness and the spires (or other vortex generating devices) must be designed to produce the type of velocity profile (the correct α value of the power-law velocity profile) and the type of turbulence similar to that encountered by the prototype structure.

Typically, the model tested in a wind tunnel is placed on a turntable so that it can be studied for winds from different directions. To simulate the wind field correctly, the model should include not only the particular structure to be tested but also all neighboring structures and terrain features—all constructed to the same scale ratio.

To maintain a constant pressure (zero pressure gradient) along tunnel test sections, the cross-sectional area of a tunnel test section must be slightly increased in the direction of wind. The increasing cross section causes a decreasing wind velocity and an increasing pressure in

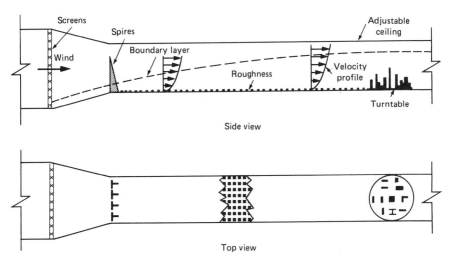

Figure 6.9 Components of boundary-layer-wind tunnel test section.

the wind direction, compensating for the pressure decrease caused by friction. Normally, the area increase is accomplished by using an adjustable ceiling for the tunnel test section. The ceiling slope is adjusted to produce a zero pressure gradient in the wind direction

6.5 MAJOR COMPONENTS OF WIND TUNNELS

As can be seen from Figures 6.1, 6.2, and 6.9, major components of a wind tunnel include fan, test section, nozzle, diffuser, honeycomb, flow straighteners, guide vanes, screens, turntable, spires, and roughness elements on floor.

The fan is needed for all types of wind tunnels. Normally, an adjustable-speed fan is used for controlling wind speeds. The test section is where the model is tested and where the atmospheric boundary layer is simulated. The model is always placed near the downstream end of the test section where the boundary layer thickness is a maximum.

The flow straighteners are large radial vanes placed immediately behind the fan (see Figure 6.2) for straightening the twisting streamlines (helical flow) of air generated by the rotation of the fan. Additional straightening is produced by the honeycomb (see same figure). Curved guide vanes are placed in all bends to prevent flow separation and to reduce energy loss. A set of screens with decreasing mesh size in the flow direction is placed upstream of the test section. The screens help to distribute the velocity uniformly across and reduce the turbulence in the wind. The screens should not be too fine (about 90% porosity), or they will cause large energy loss and velocity reduction.

The converging segment between the screens and the test section provides a smooth transition of the tunnel to a smaller cross section. The diffuser downstream from the test section provides a gradual expansion of flow without causing flow separation and excessive energy loss.

The turntable is round and it has a diameter slightly smaller than the width of the test section. It is normally turned by an electric motor. The turntable is covered with the structure to be tested and neighboring structures. Additional structures and/or topographical features must be placed upstream of the turntable to simulate the upwind structures in the prototype. Further upstream, the floor must be covered with roughness elements to generate and maintain a turbulent boundary layer. The roughness elements are normally cubic elements attached to the tunnel floor.

6.6 SIMILARITY PARAMETERS AND WIND TUNNEL TESTS

According to the similitude theory normally introduced in fluid mechanics, all model tests must be conducted under geometric, kinematic, and dynamic similarities.

Geometrical similarity requires that the shape of the model, including the structure to be tested and its surrounding structures and topographical features, must be the same as that of the prototype. Kinematic similarity means the velocity field and the streamline pattern must be similar. Finally, dynamic similarity means the pressure distribution and the forces generated by the wind must be similar. Note that these three types of similarity are not independent from each other. For instance, there cannot be kinematic and dynamic similarities without geometrical similarity. Also, dynamic and kinematic similarities are inseparable. Once the flow is kinematically similar, it must also be dynamically similar, and vice versa. Therefore, the same criteria (dimensionless parameters) that cause kinematic similarity also insure dynamic similarity.

Similarity of Approaching Flows

A prerequisite for kinematic similarity between models and prototypes is that wind tunnel tests be conducted using an approaching flow (free-stream velocity) similar to the atmospheric boundary-layer flow encountered by the prototypes. This means both the mean velocity and turbulence characteristics must be similar for models as for prototypes. To attain such a similarity, the power-law coefficient α and the relative intensity of turbulence I_r for the model flow must be the same as for the prototype, and the blockage ratio must be less than approximately 0.1. In addition, the length scales H/δ and H/L_u must be the same for the model as for the prototype, where H is a characteristic length of the structure such as its height, δ is the boundary-layer thickness, and L_u is the length scale of the turbulence defined by Eq. 3.23.

To make H/δ the same for models as for prototypes places a restriction on the size of wind tunnels. For instance, to simulate a 500-m atmospheric boundary layer in conjunction with a 100-m height building, if the model height is 0.4 m, the boundary-layer thickness in the tunnel must be $5 \times 0.4 = 2.0$ m. To use a smaller wind tunnel will not enable correct simulation of the flow. To use a much smaller model to reduce the size of the wind will not work either because this will make it difficult to simulate correctly the geometries of structures and topographical features, and difficult to attain sufficiently high Reynolds number. This explains why boundary-layer tunnels are normally huge.

Other Similarity Parameters for Stationary Structures

Kinematic similarity means more than simulating the correct approaching flow. Certain other dimensionless parameters also may have to be made identical for the model flow as for the prototype flow before the two flows can be considered kinematically similar. This is discussed next.

From fluid mechanics, for flow around a stationary object the pressure p (above ambient) at any point is a function of the velocity of the flow V, the density of the fluid ρ, the dynamic viscosity μ, the bulk modulus of elasticity E_b, and a characteristic length of the body L, namely,

$$p = \Phi(V, \rho, \mu, E_b, L) \tag{6.1}$$

where Φ is an arbitrary function.

Application of dimensional analysis to Eq. 6.1 yields

$$\frac{p}{\frac{1}{2}\rho V^2} = \Psi\left(\frac{\rho V L}{\mu}, \frac{V}{\sqrt{E_b/\rho}}\right) \quad (6.2)$$

where Ψ is another arbitrary function.

Knowing that the three parameters in Eq. 6.2 are, from left to right, the pressure coefficient C_p, the Reynolds number Re, and the Mach number Ma, the equation becomes

$$C_p = \Psi(\text{Re}, \text{Ma}) \quad (6.3)$$

Equation 6.3 shows that the pressure coefficient at any location on a stationary structure is a function of the Reynolds number Re and the Mach number Ma. However, for Mach number less than approximately one-third (i.e., for incompressible flow), the effect of Mach number is negligible, and Eq. 6.3 reduces to

$$C_p = \Psi(\text{Re}) \quad (6.4)$$

which shows that at low speed (Ma < 1/3) the pressure coefficient around a structure is a function of the Reynolds number only. Since the Mach number of the fastest natural winds such as generated by a tornado is less than one-third, the Mach number does not affect wind loads on prototype structures. As long as the Mach number for the model is also below one-third, kinematic and dynamic similarities for stationary structures can be assured by making the Reynolds number for the model the same as for the prototype.

It is a known fact that for turbulent flow around noncurved objects (i.e., structures that have no curved or round surfaces, such as a rectangular building or a billboard), flow separation occurs at the same places—the corners and edges of such structures—for a wide range of wind speeds (Reynolds number). In such a situation, kinematic and dynamic similarities prevail even if the Reynolds number is not the same for the model as for the prototype. Therefore, for noncurved objects, Eq. 6.4 further reduces to

$$C_p = \text{constant} \quad (6.5)$$

$$(C_p)_m = (C_p)_p \quad (6.6)$$

Sec. 6.6   Similarity Parameters and Wind Tunnel Tests

or

$$\left(\frac{p}{\frac{1}{2}\rho V^2}\right)_m = \left(\frac{p}{\frac{1}{2}\rho V^2}\right)_p \tag{6.7}$$

where the subscripts m and p outside the parentheses represent model and prototype, respectively.

Scaling Laws

Assuming that the density of the air in a model test is identical to that of the prototype wind, Eq. 6.7 reduces to

$$\frac{p_m}{p_p} = \left(\frac{V_m}{V_p}\right)^2 \tag{6.8}$$

Since force F is equal to pressure times area and since area is proportional to the square of the characteristic length L, Eq. 6.8 results in

$$\frac{F_m}{F_p} = \frac{p_m L_m^2}{p_p L_p^2} = \left(\frac{L_m}{L_p}\right)^2 \left(\frac{V_m}{V_p}\right)^2 \tag{6.9}$$

where F_m and F_p are, respectively, the forces on the model and the prototype buildings, and L_m/L_p is the scale ratio.

In addition, kinematic similarity makes it possible to relate model velocity and acceleration to prototype velocity and acceleration as follows:

$$\frac{V_{m1}}{V_{m2}} = \frac{V_{p1}}{V_{p2}} \tag{6.10}$$

and

$$\frac{a_{m1}}{a_{m2}} = \frac{a_{p1}}{a_{p2}} \tag{6.11}$$

where V_{m1} and V_{m2} are velocities at two arbitrary points 1 and 2 in the model, and V_{p1} and V_{p2} are the velocities at corresponding points in the prototype flow. A similar interpretation pertains to the acceleration, a.

Equations 6.8–6.11 are the scaling laws for predicting prototype behavior from model test results. They hold for both static and dynamic loads as long as the structure is stationary (not vibrating). An example is given next to bring out some salient features of wind tunnel modeling.

Suppose a model of a scale ratio $L_m : L_p = 1 : 100$ is tested in a wind tunnel. Making the Reynolds number identical for the model as for the prototype yields

$$\frac{\rho_m V_m L_m}{\mu_m} = \frac{\rho_p V_p L_p}{\mu_p} \qquad (6.12)$$

For air in ordinary wind tunnel, $\rho_m \approx \rho_p$ and $\mu_m \approx \mu_p$. Consequently,

$$\frac{V_m}{V_p} = \frac{L_p}{L_m} = 100$$

The foregoing calculation shows that to maintain the same Reynolds number, the wind speed for the wind tunnel test must be 100 times that of the prototype velocity. This means if we wish to determine the wind load on the prototype generated by a 50-m/s wind, the velocity in the wind tunnel must be $V_m = 50 \times 100 = 5000$ m/s. Not only is this velocity impractically high and hard to obtain in a wind tunnel, even if it could be obtained the test results would be totally meaningless because the flow in the tunnel would be hypersonic. This violates the assumption that the Mach number can be neglected, which is correct only for low-speed flow (Mach $< 1/3$). The only way to make the Reynolds number the same for the wind tunnel test as for the prototype wind without exceeding Mach number equal to one-third is to use a pressurized wind tunnel in which the air has very high density. Knowing that only minor changes of dynamic viscosity of air take place when air is compressed, the Reynolds number of this wind tunnel model test can be increased 100 times by compressing the atmospheric air 100 times. Once the air in the tunnel is compressed to this level, Reynolds number similarity can be satisfied by using the same velocity encountered by the prototype (50 m/s) to test the model. This explains why pressurized wind tunnels are needed for certain tests.

Fortunately, for most noncurved structures such as ordinary buildings, it is unnecessary to use the prototype Reynolds number for wind tunnel tests. As long as the Reynolds number is not too small (at least 10^4), the flow around a model will be turbulent, and kinematic and dynamic similarities will prevail even if the model Reynolds number is much smaller than the prototype Reynolds number. This provides great flexibility in the selection of wind tunnel test speeds. The speed used for structural testing is usually between 10 and 20 m/s. For speeds smaller than 10 m/s, it is difficult to get accurate readings from instruments such as pressure transducers and Pitot tubes.

Modeling Vibrating Structures

The similarity parameters for vibrating structures include all the parameters for stationary structures plus some extra parameters pertaining to vibrating structures. As soon as a structure starts to vibrate, the mass, the damping, and the elastic property (stiffness) of the structure play a role in the vibration. They must be incorporated into the similarity parameters. Mass similarity requires that the density of the fluid, ρ, to the bulk density of the structure, ρ_b, remains constant for the model as for the prototype. Damping similarity requires that the damping ratio, ζ, or the logarithmic decrement, d, discussed in the Appendix, remains constant for the model as for the prototype. Stiffness similarity requires that the Cauchy number, Ca, defined by Eq. 6.13, remain constant for the model as for the prototype.

$$Ca \text{ (Cauchy number)} = \frac{E_{\text{eff}}}{\rho V^2} \quad (6.13)$$

where E_{eff} is the effective Young's modulus of the structure.

Three types of vibration models are used in wind tunnels, they are separately discussed next.

Replica models. Replica models are full aeroelastic models. They require geometrical scaling of all dimensions and the use of the appropriate material properties. In situations where vibration is dominated by the elastic forces of structures, replica models should be based on the Cauchy number given in Eq. 6.13, except that the Young's modulus E instead of the effective Young's modulus E_{eff} should be used.

Structures that can be tested by using the Cauchy number criterion include chimneys, stacks, cooling towers, and tubular structures—all having elastic properties concentrated along their shells. From Eq. 6.13, if the air density ρ and the wind velocity V are kept the same for model as for prototype, the same material for the prototype can be used for the model.

Equivalent models. Equivalent models use a nonstructural skin to maintain the same geometry of the prototype and a structural core that simulates certain key properties of the prototype system, such as the mass and stiffness matrices. Such models are mechanical analogs of the prototypes.

A simple equivalent model often used for studying the vibration of buildings is a rigid model with springs and damping devices attached to its base. The test yields information on the fundamental-mode vibration and acceleration of the building top as well as on the dynamic forces and moments at the base. By varying the spring constant and the damping force, the dynamic behavior of the building as a function of structural stiffness and damping can be evaluated. The damping is normally provided and controlled by electromagnetic forces. Additional similarity parameters that must be satisfied include

1. The frequency ratio $(n_1)_x/(n_1)_y$, where $(n_1)_x$ and $(n_1)_y$ are the fundamental-mode (first-mode) natural frequencies of the building about the two axes of the building, x and y.
2. The logarithmic decrement d discussed in the Appendix.
3. The density ratio ρ_b/ρ, where ρ_b is the bulk density of the building and ρ is the air density.
4. The reduced velocity $V/(Bn_1)$, where V is the free-stream velocity, B is the breadth of the building in the across-wind direction, and n_1 is the natural frequency, either $(n_1)_x$ or $(n_1)_y$.

The first three parameters listed can be satisfied by adjusting the springs, the damping, and the mass of the model, respectively. The reduced velocity is used to determine the speed at which the model should be tested. For example, if the x-component of the natural frequencies of a building is 0.1 Hz and 10 Hz, respectively, for the prototype and the model, and if the scale ratio is 1 : 500, having the reduced velocity the same for the model as for the prototype yields

$$\frac{V_m}{V_p} = \frac{B_m(n_1)_m}{B_p(n_1)_p} = \frac{1}{500} \times \frac{10}{0.1} = \frac{1}{5}$$

This means the model test must be conducted at one-fifth of the prototype velocity.

Suppose we wish to predict the peak roof-level vibration of the prototype at a wind speed of 50 m/s. By testing the model at one-fifth of this wind speed (10 m/s), we found that the peak roof-level vibration of the model is 1 mm. Geometric similarity requires that

$$\frac{\hat{X}_p}{\hat{X}_m} = \frac{L_p}{L_m} = 500 \quad \text{or} \quad X_p = 500 \times 1 \text{ mm} = 0.5 \text{ m}$$

This means the peak roof-level vibration of the prototype building under 50 m/s wind will be 0.5 m. The foregoing approach can be used to determine the vibration of the prototype at various wind speeds.

A more sophisticated equivalent model is sometimes used for studying the vibration of tall, flexible buildings. The model uses several lumped mass systems to simulate the properties of each building.

Section models. For slender, linelike structures such as long-span bridges, stacks, cables, and so on, their dynamic properties can often be studied by testing segments of the structures—the section model. Section models allows the use of scale ratios many times greater than that for whole-structure models. They can be either stationary or moving. Moving section models are either driven by external devices, or mounted on springs and plucked (transient excitation) or allowed to vibrate naturally, to determine the effect of structural motion on the aerodynamic forces—the aeroelastic effect.

Gravity Effect on Vibration

The vibration of certain structures such as suspension bridges is strongly affected by gravity or the weight of the structures. Tests of such models must correctly simulate the effect of gravity. The gravity effect can be correctly simulated if we make the Froude number $V/\sqrt{gL}$ the same for the model as for the prototype, where V is the free-stream wind speed, g is gravitational acceleration, and L is a characteristic length.

By making the Froude number identical for the model as for the prototype,

$$\frac{V_m}{V_p} = \left(\frac{L_m}{L_p}\right)^{1/2} \qquad (6.14)$$

Eq. 6.14 shows that with a scale ratio of 1 : 100, the wind speed in the model test must be one-tenth of the prototype wind speed. Even smaller velocities will be required if the scale ratio is smaller than 1 : 100. Such small velocities can be troublesome for wind tunnel measurements.

Note that the Froude number is also important to the study of water-wave forces generated on structures such as an off-shore platform.

Modeling Internal Pressure

The internal pressure of a building, both the steady and fluctuating pressures, can be calculated more easily from theory than from wind tunnel tests. Wind tunnel testing of internal pressure is complicated by the fact that the leakage flow is laminar through small cracks and turbulent through large cracks. It is practically impossible to model such leakage flow correctly. Furthermore, damping of internal pressure fluctuations is caused by a combination of air leakage, building cladding flexibility and friction in materials. It is difficult to simulate these damping properties in wind tunnel tests. Consequently, wind tunnel modeling of building internal pressure is not recommended at this stage. Future advancement in the state of the art in wind tunnel tests may make internal pressure more amenable to modeling.

More about scaling laws and other aspects of wind tunnel tests can be found in many publications such as Cermak and Sadeh (1971) and an ASCE Manual of Practice (1987). A workshop proceedings (Reinhold 1982) also contains a wealth of information on wind tunnel modeling of structures.

6.7 WIND TUNNEL INSTRUMENTATION

Because most of the instruments used in wind tunnel tests are standard equipment discussed in many fluid mechanics texts, only a brief discussion of certain salient features of each important instrument will be given herein.

Pitot Tube

Pitot tube is the basic instrument used for measuring wind speed in a wind tunnel. It is based on the principle of conversion of kinetic energy to pressure at a stagnation point—the tip of the Pitot tube. The pressure differential sensed by the tube is proportional to the square of the velocity. This instrument is accurate, reliable, convenient, and economical. Furthermore, it does not require calibration. However, the Pitot tube is inaccurate at low speeds (about less than 5 m/s) and unsuitable for measuring turbulence.

Hot-Wire Anemometer

The sensing element of a hot-wire anemometer is a fine wire made of tungsten, platinum, or a special alloy. The wire is finer than human hair, and its length is only about 1 mm. The two ends of the wire are welded to two pointed electrodes (support needles) connected to a source of electricity. The turbulence in the wind causes changes of heat transfer from the wire, which in turn causes the resistance of the wire to fluctuate. The electronic circuit automatically adjusts the current going through the wire to keep the wire at constant temperature. Consequently, the velocity fluctuations (turbulence) can be determined from the fluctuations of the current through the wire. A variant of the hot-wire anemometer is the hot-film anemometer. The sensing element of a hot film is a coated metal film laid over a tiny glass wire. The rest are the same as for hot wires. The device is more robust than the hot wires and hence can be used not only in air but also in water and contaminated environments.

Hot-wire or hot-film anemometers can be used to measure both mean velocity and turbulence. They can measure rapid changes of velocities with frequency response higher than 1 kHz. Due to the small size of its sensing element, the velocity measured by a hot wire is often considered as the **point velocity**. Calibrations of hot wires are done by using a Pitot tube placed alongside a hot wire in a wind tunnel having approximately a uniform flow.

Manometers

Manometers are the standard equipment for measuring mean (time-averaged) pressure and for calibrating pressure transducers. Like Pitot tubes, manometers are accurate, reliable, and economical, and do not require calibration.

Pressure Transducers

Pressure transducers can measure both mean and fluctuating pressures. Their output are electrical signals which can be recorded and analyzed by electronic equipment including digital computers.

The sensing element of a pressure transducer is usually a small diaphragm whose deflection by air pressure causes a voltage proportional to the pressure. The voltage is measured and analyzed by electronic equipment such as a computer. For the diaphragm to generate a voltage, it must be connected to different elements such as a strain gage, a differential transformer, a capacitor, or a piezoelectric crystal. These devices generate a signal proportional to the displacement of the diaphragm. Many different types of pressure transducers can be used in wind tunnel tests. Some measure absolute pressure and some measure differential pressure. The latter is more sensitive and hence more suitable for use at low speeds.

Other Sensors

Many other transducers (sensors) may be needed in a wind tunnel study. These include strain gages for measuring strain, accelerometers for measuring the acceleration of models, and so on. They are standard sensors familiar to most structural engineers and hence not explained here.

Data Acquisition Systems

Modern data acquisition systems for wind tunnel tests consist of on-line processing of data by digital computers. Many mini- and microcomputers equipped with an analog-to-digital converter can perform such duties. The computer records the signals from various transducers, analyzes the signals, and prints or plots the results in desired forms. Such systems have brought great convenience to wind tunnel testing.

REFERENCES

ASCE Manual of Practice (1987). *Wind Tunnel Model Studies of Buildings and Structures,* Manuals and Reports on Engineering Practice No. 67, American Society of Civil Engineers, New York.

CERMAK, J. E. AND SADEH, W. Z. (1971). "Wind-Tunnel Simulation of Wind Loading on Structures," Reprint 1417 ASCE National Structural Engineering Meeting, Baltimore, Maryland.

REINHOLD, T. A. (ed.) (1982). *Wind Tunnel Modeling for Civil Engineering Applications,* Proceedings of International Workshop, Gaithersburg, Maryland, Cambridge University Press, New York.

ROSHKO, A. (1961). "Experiments on the Flow Past a Circular Cylinder at Very High Reynolds Number," *Journal of Fluid Mechanics,* 10, 345–356.

7

Building Codes and Standards[1]

7.1 INTRODUCTION

Building codes are detailed regulations (laws or ordinances) that provide minimum standards and necessary guidelines for the design, construction, alteration, repair, use, maintenance, and even demolition of buildings. In the United States, the federal government does not issue building codes; cities, counties, and some states have their own codes. It has been estimated that there are in the neighborhood of 5000 building codes in the United States (Perry 1987). However, most of these codes are based on one of three "model codes" including the Uniform Building Codes, the Standard Building Code, and the Basic Building Code. A common practice of cities, counties, and states is to adopt one of the model codes, with modifications of specific items to suit local interest. Some large cities such as New York City and Chicago and certain states such as North Carolina and Florida have created their own codes governing wind-resistant construction.

---

[1] The information provided in this chapter on specific codes and standards is that known prior to the publication of this book. Due to frequent revisions of building codes and standards, readers seeking updated information on standards and codes should consult the latest sources.

173

Building standards are more narrowly focused than building codes. Each of them provides the standards on a single aspect of building activities such as the design loads or testing of materials used in construction. They are often recognized by the building codes and become parts of the codes. For instance, beginning in the mid-1980s, all three model codes in the United States recognize the ANSI Standard A58.1-1982 entitled "Minimum Design Loads for Buildings and Other Structures." Because codes contain certain standards, in the literature the terms "codes" and "standards" are often used interchangeably.

The purpose of all building codes and standards is to ensure and enhance the safety, health, and welfare of the public. While safety is often the prime concern of building codes and standards, a careful balance is maintained between safety and economics so that the public need not pay exorbitant costs for buildings. To design exceedingly costly buildings to achieve extraordinary safety is generally not regarded as being in the best interest of the owner and the public. Consequently, building codes and standards normally claim to provide only the minimum criteria required for public safety. Any owner willing to pay a higher price to achieve a higher degree of safety can always elect to exceed the requirements of the codes and standards.

Model codes and standards are revised every two to five years to keep pace with new knowledge and technological innovation. However, cities, counties, and states may not wish to update their codes as often. Therefore, it is not uncommon for a city or a county to use an obsolete version or edition of a model code. This should not be used as an excuse to remain obsolete.

Historically, building codes and standards become increasingly complicated as the state of the art becomes more sophisticated. For instance, prior to 1972, codes and standards in the United States used a simple pressure map to specify the basic design wind pressure. The nation was divided into seven wind-pressure zones following state or county boundaries, with values of basic wind pressure varying from 20 to 50 psf at 5-psf increments. This basic wind pressure was to be applied to the first 30-ft (9.15-m) height of rectangular buildings. Higher values for larger heights were specified in a table. ANSI A58.1-1972 drastically changed this simplistic approach to a much more sophisticated method that forms the basis of modern codes and standards. The new approach, based on the probability concept and modern knowledge in wind engineering, will be discussed next.

7.2 ANSI STANDARD

Brief History

ANSI stands for American National Standards Institute, formerly known as the American Standards Association (ASA). ANSI has issued many standards unrelated to buildings. The ANSI standard for loads (wind load and other loads included) on buildings and structures is ANSI A58.1. It was first published in 1945, and subsequently revised in 1955, 1972, 1982, 1988, and so on. ANSI policy requires that the standard be reevaluated once every five years, with actions taken to reaffirm, revise, or withdraw the standard. Traditionally, the secretariat of the ANSI A58 Committee is a representative of the National Bureau of Standards (NBS), an agency whose name was changed to National Institute of Standards and Technology (NIST) in 1989.

ANSI standards are generated by a consensus process involving the development and approval of each standard by a committee. The committee members are drawn from different sectors of the society interested in the standard, such as consulting engineers, manufacturers, government agencies, and university researchers. An elaborate process to reach a consensus on each standard, including ways to resolve negative votes, is followed. In 1977, the American Society of Civil Engineers (ASCE) received accreditation from ANSI to administer (write and revise) ANSI standards related to civil engineering. As a result, the ANSI A58 Committee was transferred to ASCE in 1985 without change of its name or function.

The 1982 edition ANSI A58.1 Standard, abbreviated as ANSI-82, is entitled "Minimum Design Loads for Buildings and Other Structures." The 1988 edition is very similar to the 1982 edition, except for minor revisions. More drastic changes are expected in the next edition (Mehta 1987). The 1988 edition is referred to by ASCE as "ASCE Standard 7-88" and is referenced in this book as "ANSI/ASCE-7-1988." More about the history and the operational rules of ANSI A58 can be found in Jones (1987).

General Approach

The ANSI Standards (both the 1982 and the 1988 editions) cover all types of loads, including dead load D, earthquake load E, live loads L_r and L (respectively, for the roof and for other parts such as the floors), rain load R, snow load S, wind load W, and other possible loads such as due to the weight and the lateral pressure of soil and water in soil, earth settlement, thermal expansion, moisture change, creep in component materials, ponding, and so on.

For designs based on allowable stress instead of strength, the designer is required to use the worst load combination of the following four:

$$D + 0.75[L + (L_r \text{ or } R \text{ or } S) + (E \text{ or } W)]$$
$$D + 0.75[L + (L_r \text{ or } R \text{ or } S) + T]$$
$$D + 0.75[(E \text{ or } W) + T]$$
$$D + 0.66[L + (L_r \text{ or } R \text{ or } S) + (E \text{ or } W) + T]$$

where T is the self-straining load arising from contraction or expansion due to temperature change, shrinkage, moisture change, creep in component materials, differential settlement, or combination thereof. The factors 0.75 and 0.66 are "probability factors" accounting for the likelihood of having two or more transition loads occurring concurrently. When there is only one transition load, such as wind load or earthquake load, the factor is 1.0. A slightly different combination of loads is used for designs based on the strength of materials.

Procedure for Wind Load Determination

The ANSI procedure for calculating wind loads involves the following:

1. Determine the exposure categories. Exposure A is for centers of large cities; B is for urban, suburban, or wooded areas; C is for open terrain; and D is for unobstructed beach areas with wind coming from large bodies of water such as the sea or a large lake.
2. Determine the basic wind speed V from the map given in Figure 7.1. The basic wind speed is the fastest-mile wind corresponding to a 50-year return period (0.02 annual probability) at 10-m (33-ft) height above ground of a flat terrain of exposure category C. Values of wind speed given in this map range from 70 mph (31.3 m/s) such as for St. Louis, Missouri, to 110 mph (49.2 m/s) such as for Miami, Florida.
3. Multiply the velocity V by an importance factor I which differs for different kinds of buildings. For instance, for ordinary buildings away from hurricane oceanline, $I = 1.00$; for essential buildings such as hospitals within 100 miles (161 km) of hurricane oceanline,

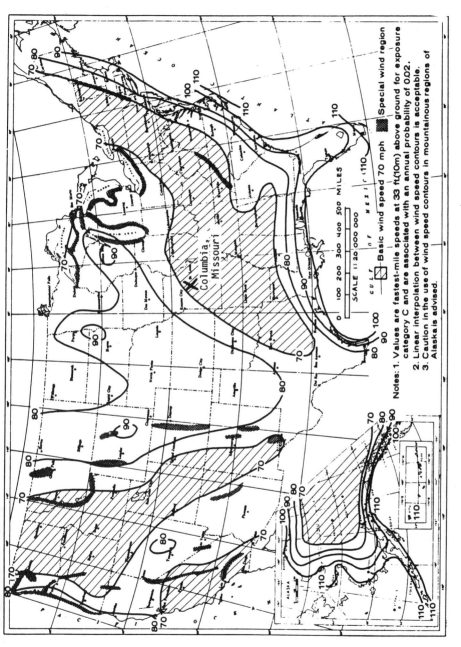

Figure 7.1 Basic wind speed map of ANSI Standard A58.1. (Courtesy of American National Standard Institute.)

$I = 1.11$; and for buildings with low hazard to human life such as agricultural buildings in nonhurricane regions, $I = 0.95$. The importance factor converts the basic wind speed determined from a 50-year return period for ordinary buildings and structures to that of a 100-year recurrence wind for essential structures and to a 25-year wind for low-hazard structures.

4. Calculate the velocity pressure at height z from

$$q = 0.00256 K_z (IV)^2 \tag{7.1}$$

where V is in miles per hour and q is in pounds per square foot. The factor K_z is called the exposure coefficient, which depends on the exposure category and the height. Values of K_z are provided by a table in the ANSI Standard.

5. Determine whether the building is flexible. A flexible building is defined as having a height to width* greater than 5, or having a natural frequency less than 1 Hz. Otherwise, it is considered a rigid building.

6. Determine the gust response factor G. For rigid structures, the gust response factor is determined from a table based on the following equation,

$$G = 0.65 + 8.58 D_o^{1/2} \left(\frac{30}{z}\right)^\alpha \tag{7.2}$$

where z is the height in feet, α is the power-law exponent, and D_o is the surface drag coefficient discussed in Chapter 3. For flexible structures, G is determined from any rational method such as given in the ANSI Appendix, or the Solari's method described in Chapter 5.

7. Determine the pressure coefficient C_p and/or the force coefficient C_f. ANSI lists the pressure coefficients and the force coefficients for various types of buildings and structures in various tables and graphs. Some of the pressure coefficients given in ANSI are combined with the corresponding gust response factor. For instance, values of GC_{pi} are given for internal pressure under various conditions.

8. The wind pressure at any point on the building is then computed by multiplying the velocity pressure calculated from Eq. 7.1 to the

---

* **Width** is the smaller of the two horizontal dimensions of a rectangular building.

product of the corresponding gust response factor and the pressure or force coefficient; for example,

$$p = qGC_p \qquad (7.3)$$

9. Once p on any building parts such as the windward wall is calculated for different heights, the resultant wind load distribution on each part of the structure can be determined.

In lieu of the foregoing procedure to calculate wind load, ANSI A58.1 also allows the wind load of any structure to be determined from properly conducted wind tunnel tests or similar model tests employing fluids other than air. Furthermore, in situations where the building has unusual geometry or response characteristics, or at sites where channeling effect or buffeting in the wake of upwind obstructions may warrant special consideration, the designers are asked either to find pertinent information in recognized literature or to use wind tunnel studies to determine the load and deflection. For more details about the standard, one should refer to the latest issue of the standard. A guide exists to the use of ANSI A58.1-1982 (Mehta 1988).

Comments on ANSI A58.1

ANSI A58.1-88 and its earlier versions do not consider the problem of tornadic wind, do not make use of the directional effect of high winds, do not show how to calculate the cross-wind response of buildings and structures, do not contain a method to estimate the amplification of wind speed by hills or escarpment, and do not provide adequate treatment of internal pressure. All these problems have been considered in earlier chapters of this book. It is anticipated that future revisions of ANSI will address these problems (Mehta 1987).

7.3 MODEL CODES

The three current model codes of the United States are the Uniform Building Code (UBC) promulgated by the International Conference of Building Officials (ICBO), the Standard Building Code (SBC) by the Southern Building Code Congress International (SBCCI), and the Basic/National Building Code by the Building Officials and Code Administrators International, Inc. (BOCA). They are normally referred to as the **UBC or ICBO Code,** the **SBC or Standard Code,** and the **Basic or BOCA Code,** respectively. Different areas of the United States use different codes. Generally, the UBC is used in most of the western states plus Hawaii and Alaska, the Standard Code covers most of the southeastern states (south of Missouri and east of Texas including Texas), and the BOCA covers most of the northeastern states. Note that much of the rural areas in the United States do not have building codes.

Uniform Building Code

The 1982 version of the UBC code (UBC-82) represents a drastic departure from its earlier versions which were based on the obsolete wind pressure maps of ANSI-55. The 1985 version (UBC-85) is a blend of UBC-82, ANSI-72, and ANSI-82.

UBC-85 allows buildings lower than 400 ft (122 m) be designed either by a simple method (Method 2) that applies a constant pressure on the projected area of the building to determine horizontal wind load, or by a more sophisticated method (Method 1) that is similar to ANSI-72 and ANSI-82. In a significant departure from ANSI, Method 2 of UBC-85 uses only two exposure categories: B (for rough terrain and urban area) and C (for smooth open areas). For structures sensitive to dynamic effects or for structures higher than 400 ft (122 m), UBC-85 requires that design be according to approved national standards—presumably ANSI A58.1. UBC-88 is an overhaul of UBC-85 containing certain improvements such as clarifying what Exposure C means and adding some subsections, one on open-frame towers and another on open buildings (Drake 1987).

Basic/National Building Code

The wind load provisions of BOCA-87 are a drastic revision of BOCA-84. It is based on ANSI-82 except that certain provisions of ANSI-82 were polished by BOCA to make it convenient for the user. There is no substantial difference between the wind load provisions of BOCA-87 and ANSI-82.

Standard Building Code (Southern Code)

For buildings of a height less than 60 ft (18.3 m), the wind load provisions of SBC-86, 87, and 88 are essentially the same as that given in a manual of the Metal Building Manufacturers Association (MBMA 1986). An extensive set of pressure coefficients, based on a wind tunnel study conducted at the University of Western Ontario, is given for the main wind force resisting systems. The maximum (peak) pressure coefficients used in MBMA and SBC are 80% of those measured in this wind tunnel study. In contrast, ANSI-82 and 88 use 100% of the peak values determined from the same wind tunnel study. For buildings higher than 60 ft (18.3 m), SBC-87 permits the use of ANSI-82 as an alternative.

Some fundamental differences in format exist between ANSI-82/ 88 and SBC-86, 87, and 88. For instance, for the calculation of the wind load on the main wind force resisting systems for buildings higher than 60 ft (18.3 m), ANSI requires that the gust response factor G and the exposure factor K_z be separately determined. In contrast, these two factors are combined with the velocity pressure in SBC. Also, while ANSI-82 and 88 multiplies the basic wind velocity by an importance factor I, the SBC Code multiplies the pressure calculated by the square of I, which it calls the use factor. A detailed comparison between ANSI and SBC is given in Vognild (1987).

Comparisons of the wind load provisions of ANSI to those of all the model codes are given in McBean (1987) and Perry (1987).

7.4 INTERNATIONAL STANDARDS

With increasing world trade of goods and services, there is an increased need for American engineers and architects to know international codes and standards. While ANSI still uses English units such as miles per hour and pounds per square foot, practically all other nations, including Great Britain, Canada, and Australia, use SI units. A brief description of certain important international standards on wind load is given next.

Canadian Standard

The evolvement of ANSI-72 and later editions have been strongly influenced by the National Building Code of Canada (NBCC). There are more similarities than dissimilarities between the two standards. Some salient features of the Canadian Code are as follows:

In contrast to ANSI, which is based on the fastest-mile wind, the basic wind speed V used in NBCC is the hourly average of fastest winds which represents a much longer averaging time. Consequently, the values of the basic wind speeds used in NBCC are smaller than those for ANSI.

Another unique feature of NBCC is that it uses a smaller return period of 10 years for designing claddings and for considering structural vibration and deflection, than for designing the structural members of ordinary buildings which use a 30-year return period. In contrast, ANSI uses the same return period for cladding as for structural members.

In NBCC-85, the velocity pressure is calculated from

$$q = \frac{1}{2}\rho V^2 = 0.0027 V^2 \quad \text{(English units)}$$

$$= 0.65 V^2 \quad \text{(SI units)} \quad (7.4)$$

The design pressure is calculated from

$$p = qC_e C_g C_p \quad (7.5)$$

where C_e is an exposure factor, C_g is the **gust effect factor**, and C_p is the pressure coefficient.

Note that NBCC uses three exposure categories—A, B, and C,—with A being the smoothest terrain (open country) and C being the roughest (center of large cities). As discussed before, ANSI uses four exposure categories, with D being the smoothest and A being the roughest. Due to the use of hourly wind speed instead of the fastest mile, the "gust effect factor" used in the Canadian Standard is considerably higher than the "gust response factor" used in ANSI. For instance, NBCC gives a simple procedure for low- and medium-height buildings which uses $C_g = 2.0$ for buildings as a whole and for main structural members, and $C_g = 2.5$ for small elements including cladding. The values of the gust response factor G in ANSI are normally of the order of 1.5.

Finally, the National Research Council of Canada publishes a supplement to NBCC that gives detailed information on data and methodologies needed for wind load determination. The Supplement, although not a part of the Code, is very useful for designers.

British Standard

The British standard on wind load is issued by the British Standard Institute (BSI). Since 1970, this standard has used the modern approach of wind load determination based on return period, pressure coefficients, internal pressure, and so on. The standard, in one of its appendices, also includes a method to estimate the increase of wind speed produced by cliffs or escarpments. The British standard is supplemented by a detailed handbook (Newberry and Eaton 1974).

The basic wind speed used in the British standard is the 3-second gust at 10 m (32.8 ft) above ground in open level country, using a

50-year return period. Due to the use of gust speed, no gust effect factor or gust response factor is needed.

The British standard calculates the design wind speed V_s from the basic wind speed V using the following formula:

$$V_s = S_1 S_2 S_3 V \qquad (7.6)$$

S_1 is the "topography factor" which accounts for increase or reduction of wind speed due to local topography. For instance, the standard uses $S_1 = 1.1$ for exposed hill slopes and valleys that accelerate the wind, and uses $S_1 = 0.9$ for valleys sheltered from wind. S_2 is the "exposure factor" which depends on ground roughness, building size, and height above ground. S_3 is the "statistical factor" which accounts for the degree of safety required for a building and the life span of the building. S_3 is smaller than 1.0 for temporary structures and greater than 1.0 for structures that require greater than normal safety.

Once V_s is calculated from Eq. 7.6, the velocity pressure is determined from

$$q = 0.00256 V_s^2 \quad \text{(English units)}$$

$$= 0.613 V_s^2 \quad \text{(SI units)} \qquad (7.7)$$

The design pressure is then calculated from

$$p = q C_p \qquad (7.8)$$

In 1988, the British Standard was undergoing a major revision. More about the Canadian and British standards is discussed in Stathopoulos (1987).

Australian Standard

Because the bulk of Australia's border (coastline) except the south is threatened by devastating cyclones, Australia has been in the forefront to modernize wind load provisions of building codes and to specify wind-resistant housing construction—the deemed-to-comply provisions.

The Australian standard for wind load is issued by the Standard Association of Australia (SAA). The first major revision (modernization) comparable to ANSI-72 took place in 1971. This Australian standard, together with later minor revisions, form the basis of the following discussion.

Similar to the British Standard, the Australian Standard is based on extreme gust of 2- to 3-second duration. Four return periods are used: 50 years for all ordinary structures, 100 years for structures having postdisaster functions (e.g., hospitals and communications centers), 25 years for structures of low hazard to life and property, and 5 years for structures used only during construction such as formwork (scaffolding). The basic wind speeds corresponding to these four return periods for all major cities are listed in a table. A wind speed map is given for the entire nation. All the cyclone-prone areas use a minimum basic wind speed of 55 m/s (123 mph) for 50-year return periods and use a minimum of 60 m/s (134 mph) for 100-year return periods—all based on the gust speed.

The basic wind speed V is adjusted for terrain category and height to yield the design velocity V_z. The velocity pressure q is calculated from

$$q = 0.6 V_z^2 \qquad (7.9)$$

where q is in Pascals and V_z is in meters per second.

Finally, the design wind pressure is calculated from

$$p = qC_p \qquad (7.10)$$

In the treatment of internal pressure, the Australian Standard specifies that an openable window be considered as either open or closed in determining the most critical loading condition. This is certainly the most conservative if not the most logical approach used in treating internal pressure.

As in the case of the British Standard, the Australian Standard shows how to adjust wind speed near a cliff or escarpment.

The Australian Standard underwent major revision in 1987–88. The new standard (Australian Standard 1989) incorporated the concept of limit-states design. Two characteristic wind speeds are used for limit-states design: V_u for ultimate (collapse and overturn) and V_s for serviceability. While V_u is the gust speed corresponding to an exceedence probability of 5% in 50 years, V_s is the gust speed for an exceedence probability of 5% in any one year.

The new standard also gives the directional wind speeds, in eight directions, for major cities in the noncyclone zone. The use of directional wind speeds normally results in 80 to 85% load reduction. In regions where directional wind speed is unavailable or when designers do not wish to use this information, a directional load reduction factor of 0.90 is allowed.

The new standard requires the consideration of the acceleration of wind over slopes that are steeper than 5%. A method is given to estimate the speed-up of wind over hills and escarpments.

Finally, the new standard treats the cross-wind dynamic response of rectangular buildings, and gives an expanded list of pressure and force coefficients for various types of buildings having monoslope, pitched, curved, or multispan roofs, and for circular bins, silos, tanks, canopies, awnings, carports, and so on (Holmes, 1987).

Japanese Standard

Japanese designers use the **Building Standard Law of Japan** (Japanese Standard 1987) for the design of ordinary buildings and houses. For tall or special buildings, they use Recommendation for Design Loads on Buildings in conjunction with boundary-layer wind tunnel tests. For long-span bridges (suspension bridges and cable-stayed bridges), they use wind resistant design criteria for the Honshu-Shikoku bridges. And for transmission towers and lines, they use **Design Standards on Structures for Transmission** (JEC-127).

The approach used for ordinary buildings is similar to the 1970 Australian Standard and the "simple procedure" in ANSI-82 and 88. The wind load per unit area is determined from

$$w = Cq \qquad (7.11)$$

where C is shape factor and q is velocity pressure at height z.

For heights not exceeding 16 m (52.5 ft), q is calculated from

$$q = 0.588z^{1/2} \qquad (7.12)$$

where z is in meters and q in kilopascals.

For heights greater than 16 m,

$$q = 1.176z^{1/4} \qquad (7.13)$$

The last two equations are based on one-fourth and one-eighth power laws, respectively, and on peak gust speed profiles of the most severe typhoons encountered in Japan.

The approach for designing tall or special buildings is similar to the "detailed procedure" of ANSI-82 and 88. More about the Japanese standards can be found in Kawai (1987).

ISO Standard

In 1987, a group of international wind-load experts working under the auspices of the International Standard Organization (ISO) developed a proposed ISO Wind Load Standard. This standard sets forth the general approach to be used for wind load determination.

The wind force per unit area w (same as pressure p except that occasionally w represents skin friction or shear) is calculated from

$$w = q_r C_e C_s C_d \tag{7.14}$$

where q_r is "reference velocity pressure," C_e is "exposure factor," C_s is "aerodynamic shape factor," and C_d is "dynamic response factor."

The reference velocity pressure q_r corresponds to a 10-minute mean velocity pressure at 10-m height above an open terrain—based on a 50-year return period. The exposure factor C_e takes into account not only height and terrain roughness, but also the speed-up of wind over hills and escarpments. Values of the shape factor C_s are primarily the same as those given in the National Building Code of Canada. The ISO standard contains a detailed discussion of how to calculate the dynamic response factor C_d for different cases such as (1) small rigid structures; (2) large rigid structures; (3) along-wind response of flexible structures, (4) cross-wind, torsional, and other responses due to gust; and (5) vortex-shedding-generated cross-wind response. Some of these have been considered in previous chapters. Details of the ISO Standard are contained in Davenport (1987).

European Community Standard

With increasing cooperation and decreasing national barriers among European countries, the Commission of the European Communities (CEC) drafted a set of common codes (standards) for its member nations—the **European Codes** or **EUROCODES**.

The wind load provisions of the EUROCODES allows that structures be considered as "rigid" if they are less than 50 m high or if dynamic deflection (fluctuations) caused by wind forces is less than 10% of the quasi-static deflection.

The wind force per unit area is determined from

$$w = qGC_p \tag{7.15}$$

where q is the velocity pressure based on the design wind speed V at height z, namely,

Sec. 7.5   *Prescriptive Codes*

$$q = \frac{1}{2} \rho V_z^2 \qquad (7.16)$$

and V is calculated from

$$V = C_e V_r \qquad (7.17)$$

where C_e is "exposure factor" and V_r is "reference velocity." The reference velocity is that measured at 10 m above flat open terrains, using a 10-minute averaging time and a 50-year return period.

The exposure factor C_e accounts for the variation of wind speed with height and terrain roughness. It is determined from the power-law profile as follows:

$$C_e = b \left(\frac{z}{10}\right)^\alpha \qquad (7.18)$$

where α and b are given for three terrain categories.

Two methods are used to determine the gust response factor G—a simple method and a detailed method. The simple method is for "ordinary structures." It uses $G = 2.0$ for structures and $G = 2.5$ for components and claddings.

In the detailed method, G is calculated from

$$G = 1 + 2gI_r B_a \qquad (7.19)$$

where g is the "peak factor" (equal to 3 for whole structures and 4 for components and claddings), I_r is the relative intensity of turbulence, and B_a is the "background response factor" which depends on the width-to-height ratio of buildings. More about the European Standard is discussed in Ruscheweyh (1987).

7.5 PRESCRIPTIVE CODES

Introduction

The wind load provisions discussed in the previous sections are often referred to as the **performance standards** or **performance codes,** for they specify the expected minimum performance of structures such as the minimum wind force or pressure that a structure must be able to withstand. Building codes also contain detailed prescriptions (specifications) as to what the building materials should be and how various parts of a building must be constructed. They are often referred to in American literature as the **prescriptive codes**, or **specification codes**. Australian literature refers to them as **deemed-to-comply standards**. An example of prescriptive codes in U.S. model codes (UBC, Basic, Standard) is the provisions on how to connect the members or make joints of timber structures—the nailing schedule.

Although prescriptive codes provisions such as nailing schedules give the appearance of being unrelated to wind loads, they are closely related to building safety in high winds, especially for small buildings (homes, churches, small schools, small commercial buildings). These buildings receive little or no engineering attention. They are designed and built by nonengineers and sometimes nonarchitects, without performing structural analysis in the design. As long as building codes continue to allow such a practice, it is vitally important to have good prescriptive codes to assure building safety in high winds.

According to a 1988 report[2] of the Task Committee of Mitigation of Wind Damage, American Society of Civil Engineers, most building codes in the United States contain outdated prescriptive provisions which contribute to the widespread losses caused by hurricanes, tornadoes, and other high winds. Updating these codes by incorporating new knowledge in wind-resistant construction techniques will go a long way toward mitigating wind damage to buildings.

Two codes in the United States that have exercised leadership in providing good prescriptive provisions for wind-resistant construction are the North Carolina Uniform Residential Building Code and the South Florida Building Codes, hereafter referred to simply as the "North Carolina Code" and the "South Florida Code." At the time of writing of this book, the Southern Building Code Congress International is developing a special "deemed-to-comply" prescriptive standard for single- and multiple-family houses in high-wind regions (SBCCI 1989). The proposed standard specifies the requirements for the design and construction of both wood and masonry houses, in three different wind regions: 90, 100, and 110 mph (fastest-mile basic wind speed).

North Carolina Code

The North Carolina Code, modified in 1986, requires wind-resistant construction for coastal regions. Some of the wind-resistant

---

[2] The report was published as a set of papers in two issues of the *Journal of Aerospace Engineering,* ASCE (April and October 1989).

Sec. 7.5   Prescriptive Codes

construction provisions of the code, as discussed in (Sparks 1987), are the following:

WOOD-FRAME BUILDINGS

1. Each rafter must be anchored to the bearing plate by metal anchors or ties.
2. Near the ridge, each rafter must have a collar beam of minimum dimension of 1 in. × 6 in.
3. Every third rafter must be anchored vertically with minimum 1 in. × 6 in. member from its midpoint to the ceiling joist below.
4. Wall studs must be covered by 15/32-in. plywood sheathing which overlap the top plate and the sill, beam, or girder at least by 6 in. In lieu of the plywood system, 3/8-in. hot-dip galvanized steel rods which provide a continuous tie from the top plate down through the sill, beam, or girder may be used. The rods must be no more than 8 ft on center, and no more than 2 ft from each corner. Each end of the rod must have a washer of minimum diameter of 3 in.
5. Exposed metal connectors and nails must be hot-dip galvanized.

MASONRY BUILDINGS

1. Rafters and joists must be securely anchored to footing by 3/8-in. steel rods not more than 8 ft apart, one of which must be no more that 2 ft from each corner.
2. Sills, beams, or girders must be anchored to footing with 5/8-in. steel rods containing 10-in. hooks embedded at least 6 in. in concrete.
3. Steel and wood columns and posts, including porch columns, must be anchored with metal ties and bolts to their foundations and to the members they support.

South Florida Code

The South Florida Building Code, for the Greater Miami and Fort Lauderdale areas covering 4 million people, is primarily a specification-type code. This code is especially good at specifying how to construct masonry buildings against the strong forces of a hurricane wind and against hurricane-generated wave actions. A few typical provisions on concrete-block type construction, as mentioned in Saffir (1983), are given as follows:

1. Each concrete-block wall panel must be surrounded by reinforced concrete beams and columns. The area of each panel must not exceed 256 ft^2.
2. For buildings more than one-story high, reinforced concrete columns must exist at all corners, at intervals not to exceed 20 ft on center, adjacent to any corner opening exceeding 4 ft in width, adjacent to any wall opening exceeding 9 ft in width, and at the ends of freestanding walls exceeding 2 ft in length. Structurally designed columns may be substituted for the tie columns specified in the code.
3. For one-story residential buildings, the reinforced concrete tie beam must be anchored at intervals not to exceed 20 ft on center to the foundation or floor slab. Tie columns must not be less than 12 in. in width, with an unbraced height of not over 15 ft. The columns are reinforced with no fewer than four 5/8-in. steel vertical bars for 8-in. × 12-in. columns, and no fewer than four 3/4-in. bars for 12-in. × 12-in. columns, and so forth.

The foregoing discussion provides a glimpse of what is contained in two of the best specification or prescriptive codes in the United States. The importance of having similar wind-resistant construction provisions in all building codes in the nation and the world cannot be overemphasized.

REFERENCES

ANSI A58.1 (1982). *Minimum Design Loads for Buildings and Other Structures*, American National Standard Institute, New York.

Australian Standard (1975). *SAA Loading Code*. Part 2: *Wind Forces*, Standards Association of Australia, North Sydney.

Australian Standard (1989). *SAA Loading Code*. Part 2: *Wind Loads*, Standards Association of Australia, North Sydney.

British Standard (1970). *BSI Code of Practive, CP3*. Chapter V: *Part 2, Wind Forces*, British Standard Institute, London.

Canadian Standard (1985). *National Building Code of Canada*, National Research Council of Canada, Ottawa.

DAVENPORT, A. G. (1987). "Proposed New International (ISO) Wind Load Standard," *Proceedings of the WERC/NSF Symposium on High Winds and Building Codes*, Kansas City, Missouri, November 2–4, 1987, University of Missouri-Columbia, Engineering Extension (H. Liu, editor)., 373–388.

DRAKE, F. M. (1987). "Wind Design Provisions of the Uniform Building Code," *Proceedings of the WERC/NSF Symposium on High Winds and Building Codes,* Kansas City, Missouri, November 2–4, 1987, University of Missouri-Columbia, Engineering Extension (H. Liu, editor)., 77–83.

HOLMES, J. D. (1987). "The New Australian Standard on Wind Forces—Philosophy, Format, and Contents," *Proceedings of the WERC/NSF Symposium on High Winds and Building Codes,* Kansas City, Missouri, November 2–4, 1987, University of Missouri-Columbia, Engineering Extension (H. Liu, editor)., pp. 365–372.

Japanese Standard (1987). Building Standard Law of Japan, Building Center of Japan.

JONES E. (1987). "Procedures for Revising A58.1 Provisions," *Proceedings of the WERC/NSF Symposium on High Winds and Building Codes,* Kansas City, Missouri, November 2–4, 1987, University of Missouri-Columbia, Engineering Extension (H. Liu, editor)., 25–30.

KAWAI, H. (1987). "Wind Load Design Standards and Practice in Japan," *Proceedings of the WERC/NSF Symposium on High Winds and Building Codes,* Kansas City, Missouri, November 2–4, 1987, University of Missouri-Columbia, Engineering Extension (H. Liu, editor)., 397–404.

LIU, H. (1987). "Treatment of Internal Pressure in Building Codes," *Proceedings of the WERC/NSF Symposium on High Winds and Building Codes,* Kansas City, Missouri, November 2–4, 1987, University of Missouri-Columbia, Engineering Extension (H. Liu, editor)., 233–240.

MARSHALL, R. D. (1987). "1988 Revision of ANSI A58.1," *Proceedings of the WERC/NSF Symposium on High Winds and Building Codes,* Kansas City, Missouri, November 2–4, 1987, University of Missouri-Columbia, Engineering Extension (H. Liu, editor)., 39–45.

MBMA (1986). *Low-Rise Building System Manual,* Metal Building Manufacturers Association, Cleveland, Ohio.

MCBEAN, R. P. (1987). "Comparison of Wind Load Provisions of ANSI A58.1-1982 and Model Codes in the United States," *Proceedings of the WERC/NSF Symposium on High Winds and Building Codes,* Kansas City, Missouri, November 2–4, 1987, University of Missouri-Columbia, Engineering Extension (H. Liu, editor)., 53–60.

MCCLEUER, R. (1987). "Highlights of Wind Load Provisions of 1987 BOCA National Building Code," *Proceedings of the WERC/NSF Symposium on High Winds and Building Codes,* Kansas City, Missouri, November 2–4, 1987, University of Missouri-Columbia, Engineering Extension (H. Liu, editor)., 93–95.

MEHTA, K. C. (1987). "Wind Load Provisions of the Future," *Proceedings of the WERC/NSF Symposium on High Winds and Building Codes,* Kansas City, Missouri, November 2–4, 1987, University of Missouri-Columbia, Engineering Extension (H. Liu, editor)., 47–51.

MEHTA, K. C. (1988). *Guide to the Use of the Wind Load Provisions of ANSI A.58.1,* Institute for Disaster Research, Texas Tech University, Lubbock.

NEWBERRY, C. W. AND EATON, K. J. (1974). Wind Loading Handbook, British Building Research Establishment, London.

PERRY, D. C. (1987). "Building Codes in the United States—An Overview," *Proceedings of the WERC/NSF Symposium on High Winds and Building Codes,* Kansas City, Missouri, November 2–4, 1987, University of Missouri-Columbia, Engineering Extension (H. Liu, editor)., 65–75.

RUSCHEWEYH, H. (1987). "The Development of an European Wind Load Code," *Proceedings of the WERC/NSF Symposium on High Winds and Building Codes,* Kansas City, Missouri, November 2–4, 1987, University of Missouri-Columbia, Engineering Extension (H. Liu, editor)., 405–412.

SAFFIR, H. S. (1983). "Practical Aspects of Design for Hurricane-Resistant Structures: Wind Loadings," *Journal of Wind Engineering and Industrial Aerodynamics,* Vol. 11, 247–259.

SAFFIR, H. S. (1987). "State of Florida and South Florida Building Codes to Prevent Hurricane Damage," *Proceedings of the WERC/NSF Symposium on High Winds and Building Codes,* Kansas City, Missouri, November 2–4, 1987, University of Missouri-Columbia, Engineering Extension (H. Liu, editor)., 115–128.

SBCCI (1989). *A Deemed-to-Comply Standard for Single- and Multifamily Dwellings in High Wind Regions,* Southern Building Code Congress International, Birmingham, Alabama.

SPARKS, P. R. (1987). "The North Carolina Residential Building Code and Its Wind Load Requirements," *Proceedings of the WERC/NSF Symposium on High Winds and Building Codes,* Kansas City, Missouri, November 2–4, 1987, University of Missouri-Columbia, Engineering Extension (H. Liu, editor)., 107–114.

STATHOPOULOS, T. (1987). "Wind Load Features of Canadian and British Standards," *Proceedings of the WERC/NSF Symposium on High Winds and Building Codes,* Kansas City, Missouri, November 2–4, 1987, University of Missouri-Columbia, Engineering Extension (H. Liu, editor)., 389–396.

Supplement to NBCC (1985). National Research Council of Canada, Ottawa.

VOGNILD, R. A. (1987). "The Wind Load Path from ANSI to the Standard Building Code," *Proceedings of the WERC/NSF Symposium on High Winds and Building Codes,* Kansas City, Missouri, November 2–4, 1987, University of Missouri-Columbia, Engineering Extension (H. Liu, editor)., 85–91.

Appendix
Basic Concepts on One-Dimensional Vibration

A prerequisite for understanding the complex behavior of structural vibration is the basic concept of one-dimensional (single-degree-of-freedom) vibration of a mechanical system. Several cases of such vibration are described next.

Undamped Free Vibration

Suppose a mass m is attached to a spring having a stiffness constant k. Without damping and without forcing function, we have the following,

$$m\ddot{X} + kX = 0 \qquad (A.1)$$

where X is the displacement of the mass from its equilibrium position and $\ddot{X}$ represents the second derivative of x with respect to time which means acceleration.

The solution of Eq. A.1 is

$$X = X_O \sin(\sqrt{k/m}\ t + \phi) \qquad (A.2)$$

where X_o is the initial displacement and ϕ is the initial phase angle.

From Eq. A.2, the quantity $\sqrt{k/m}$ is the *angular velocity* or **circular frequency** of the system ω_o, namely,

$$\omega_o = 2\pi n_o = \sqrt{k/m} \qquad (A.3)$$

where n_o is the frequency in hertz.

From Eq. A.3, the **natural frequency,** that is, the frequency of vibration of the system, is

$$n_o = \frac{\omega_o}{2\pi} = \frac{\sqrt{k/m}}{2\pi} \qquad (A.4)$$

Damped Free Vibration

If damping is present in the system due to the existence of a dashpot or shock absorber, vibration is now governed by

$$m\ddot{X} + c\dot{X} + kX = 0 \qquad (A.5)$$

where c is the **damping constant** and $\dot{X}$ is the first derivative of X, which means velocity.

The solution of Eq. A.5 depends on the value of $c/(2\sqrt{km})$, which is called the **damping ratio,** ζ, namely,

$$\zeta \text{ (damping ratio)} = \frac{c}{2\sqrt{km}} = \frac{c}{c_o} \qquad (A.6)$$

Note that the denominator $2\sqrt{km}$ has been designated as c_o—the critical damping constant.

Equation A.5 has three types of solutions: **underdamped** when $\zeta < 1$ (i.e., $c < c_o$), **critically damped** when $\zeta = 1$ (i.e., $c = c_o$), and **overdamped** when $\zeta > 1$ (i.e., $c > c_o$).

In the underdamped case ($\zeta < 1$), the solution of the differential Eq. A.5 is

$$X = X_o e^{-bt} \sin(qt + \phi) \qquad (A.7)$$

where

$$b = \frac{c}{2m} \quad and \quad q = \sqrt{\frac{k}{m} - \left(\frac{c}{2m}\right)^2} \qquad (A.8)$$

Equation A.7 represents a damped oscillation as shown in Figure A.1. The amplitudes of two successive periods of oscillation, X_1 and X_2, are marked in the graph. From Eq. A.7,

Appendix   Basic Concepts on One-Dimensional Vibration   195

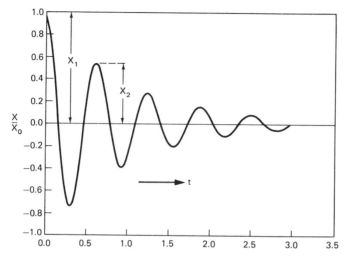

Figure A-1   Damped vibration with $\zeta < 1$.

$$\frac{X_1}{X_2} = \frac{e^{-bt}}{e^{-bt} + (2\pi/q)} = e^{2\pi b/q} \qquad (A.9)$$

Taking the natural logarithm of both sides of the equation yields the logarithmic decrement d as follows:

$$d = \ln \frac{X_1}{X_2} = \frac{2\pi b}{q} = \frac{2\pi \zeta}{\sqrt{1 - \zeta^2}} \qquad (A.10)$$

The logarithmic decrement is a dimensionless number that characterizes the rate of damping of vibration. Its value is zero when ζ is zero (undamped), and it approaches infinity when ζ approaches 1.0 (critically damped). One of the similarity parameters used in wind tunnel modeling of structural vibration is the logarithmic decrement; see Chapter 6.

Damped Vibration with Sinusoidal Forcing Function

With a sinusoidal forcing function, the vibration equation can be written as

$$m\ddot{X} + c\dot{X} + kX = F_o \sin(\omega t + \phi) \qquad (A.11)$$

where F_o is the amplitude of the forcing function, ω is the circular frequency of the forcing function, and ϕ is an arbitrary phase angle.

The general solution of the equation is

$$X = X_o \sin(\omega_o t + \phi) + e^{-bt}[C_1 \sin(qt) + C_2 \cos(qt)] \quad (A.12)$$

where

$$X_o = \frac{F_o/m}{\sqrt{(\omega_o^2 - \omega^2)^2 + (c\omega/m)}} \quad (A.13)$$

The quantities b, ω_o, q, and ϕ are as defined before. Both C_1 and C_2 are arbitrary constants that can be determined from boundary conditions.

Dropping the term with the factor e^{-bt} in Eq. A.12 for large t yields the steady-state solution:

$$X = X_o \sin(\omega_o t + \phi) \quad (A.14)$$

A **magnification factor** (also called **mechanical admittance function**) may be defined as

$$H_1 = \frac{X_o}{X'} \quad (A.15)$$

where $X' = F_o/k$. Substituting Eq. A.13 into Eq. A.15 yields

$$H_1 \text{ (magnification factor)} = \frac{1}{\sqrt{(1 - \mu^2)^2 + 2\zeta\mu^2}} \quad (A.16)$$

where $\mu = \omega/\omega_o = n/n_o$ is called the **frequency ratio** and $\zeta = c/c_o$ is the **damping ratio**.

From Eq. A.16, maximum magnification occurs when the frequency ratio μ approaches 1.0 and when the damping ratio ζ approaches zero. This can be seen from plotting Eq. A.16 in Figure A.2.

Damped Vibration with Random Forcing Function

When the forcing function is random, such as generated by the turbulence of wind, vibration is governed by

$$m\ddot{X} + c\dot{X} + kX = F(t) \quad (A.17)$$

where $F(t)$ is the random forcing function.

A random function of time is said to be **stationary** if its statistical properties such as the mean, standard deviation, spectrum, and so on do not vary with the time origin used in determining these statistical

Appendix   Basic Concepts on One-Dimensional Vibration   197

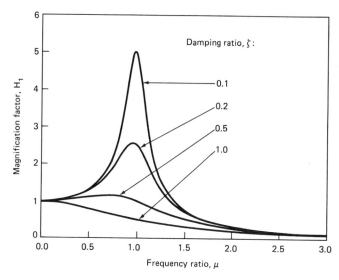

Figure A-2 Variation of magnification factor H_1 with frequency ratio μ and damping ratio ζ.

properties. Suppose $F(t)$ is a random stationary function, it can be proved from Eq. A1.17 that

$$S_X(n) = H_1^2(n) \, S_F(n) \quad (A.18)$$

where $S_X(f)$ is the power spectrum of the response (i.e., the power spectrum of the vibration X), $S_F(n)$ is the power spectrum of the forcing function $F(t)$, and H_1 is the magnification factor given by Eq. A.16.

Eq. A.18 shows that the response spectrum S_X can be calculated simply from the product of H_1 and S_F. Since S_X, S_F, and H_1 are all functions of the frequency n, Eqs. A.18 and A.16 must be used many times at various frequencies to get the response spectrum. The vibration system behaves like a nonlinear amplifier which amplifies the spectrum of the random forcing function $S_F(n)$ according to the value of H_1 at any frequency n to yield the spectrum of the response, $S_X(n)$.

Finally it is worthwhile to mention that by using Eqs. A.3 and A.6, Eq. A.17 can be written in the following alternate form:

$$\ddot{X} + 2\omega_o \zeta \dot{X} + \omega_o^2 X = \frac{F(t)}{m} \quad (A.19)$$

Vibration of Cylinders

The foregoing discussion of vibration pertains to that of a point mass. The same concept can be applied to the vibration of a cylinder held normal to wind, such as the vortex-shedding-induced vibration of a cable or the galloping of a cable-suspended bridge deck. For cylinders, Eqs. A.17 and A.19 are rewritten for a unit length of the cylinders as follows:

$$m\ddot{X} + c\dot{X} + kX = f(t) \qquad (A.20)$$

and

$$\ddot{X} + 2\omega_o\zeta\dot{X} + \omega_o^2 X = \frac{f(t)}{m} \qquad (A.21)$$

where m is the mass per unit length of the cylinder, X is the amplitude of vibration perpendicular to wind, and $f(t)$ is the corresponding force on unit length of the cylinder.

CONVERSION OF UNITS AND CONSTANTS

Length
1 m = 3.28 ft, 1 mile = 5280 ft = 1.609 km, 1 km = 1000 m, 1 m = 100 cm, 1 ft = 12 in., 1 in. = 2.54 cm

Velocity
1 knot = 1.15 mph, 1 mph 0.447 m/s = 1.467 fps, 1 m/s = 2.24 mph

Frequency
1 Hz (hertz = 1 cps)

Mass
1 kg = 2.205 lbm = 0.0685 slug, 1 slug = 32.2 lbm = 14.6 kg

Density
1 kg/m^3 = 0.001941 slug/ft^3

Force
1 N (newton) = 10^5 dynes = 0.2248 lbf, 1 lbf = 4.45 N

Pressure and Shear
1 Pa (pascal) = 1 N/m^2 = 0.0209 psf, 1 psf = 47.9 Pa, 1 kPa = 1000 Pa, 1 bar = 10^5 Pa = 2090 psf, 1 bar = 10^3 mb (millibars)

Dynamic Viscosity, μ
1 poise = 1 dyne-s/cm^2, 1 lb-s/ft^2 = 479 poises = 47.9 N-s/ft^2

Kinematic Viscosity, ν
1 stoke = 1 cm^2/s, 1 ft^2/s = 929 stokes = 0.0930 m^2/s

Properties of Air (Standard Atmospheric Condition)
p = 14.7 psia = 101.4 kPa, T = 59 °F = 15 °C
ρ = 0.00237 slug/ft^3 = 1.22 kg/m^3
μ = 3.72 × 10^{-7} lb-s/ft^2 = 1.78 × 10^{-8} kN-S/m^2

Equation of State of Perfect Gas

$$\rho = \frac{p}{RT} \quad \begin{aligned} R &= 1715 \text{ ft-lb/slug °R} \\ &= 286.8 \text{ N-m/kg °K} \end{aligned}$$

Index

Acceleration of buildings and
  structures:
  peak, 132
  rms (root-mean-square), 132
Across-wind response of structures,
  136–40
  Australian Standard method, 137
  Canadian method, 138–40
Aerodynamic instability, 110
Aeroelastic instability, 110
Aeroelasticity, 110
Aircraft damage by wind, 22
Along-wind response of structures
  (See Dynamic response of
  buildings and structures)
Ambient pressure, 70
Ambient tube, 70
Amplification factor, 57
Amplification of wind:
  by buildings, 62–65
  by topography, 55–62
Angle of attack, 101–2
Annual mean (long-term average)
  wind speed, 47
Annual series, 32

ANSI Standard on wind load,
  174–79
Arcade effect on wind, 64–65
Architectural features, effect on
  wind, 83
ASCE Standard on wind load,
  175–79
ASCE Task Committee report, 188
Atmospheric boundary layer, 41–42
  thickness, 43
Atmospheric pressure change:
  in hurricanes, 3–5
  in tornadoes, 14
Australian standard, 137–38, 183
Autocorrelation coefficient, 53
Averaging time, 46–47

Beaufort scale, 6
Blast effect of internal pressure, 89
Bora, 23
Boulder, Colorado, wind problems
  of, 23
Boundary layer, 41–42
Boundary-layer thickness, 43

Boundary-layer wind, 41
Boundary-layer wind tunnel, 78
Bridges:
  Tacoma Narrows, destruction of, 144
  torsional divergence, 121
  vibration of, 121, 144
  wind tunnel test of, 150
British Standard, 182–83
Buffeting, 83, 110, 123–24
Building codes, 173–92 (*See also* Building standards)
  ANSI A58.1, 174–79
  Basic/National (BOCA), 179–80
  definition, 173
  historical development, 174–75
  international, 181–87
  Metal Building (MBMA), 180
  model codes, 173, 179–81
  North Carolina, 188–89
  performance-type, 187–90
  prescriptive-type, 187–90
  purpose, 174
  South Florida, 189–90
  specification-type, 187–90
  Standard (Southern), 179–80
  Uniform (UBC), 179–80
Buildings:
  breadth, 76
  codes, 173–92 (*See* Building codes)
  depth, 76
  drag, 98–99
  effect on wind, 62–65
  flow around, 77
  height, 76
  length, 76
  lift, 101–2
  normal forces, 103
  overturning moment, 103–4
  pressure coefficients (*See* Pressure coefficients)
  shear, 104–6
  standards (*See* Building standards)
  vibration, wind-induced, 124, 132–34
  width, 76
Building-spawned winds, 62–65
Building standards, 173–92 (*See also* Building codes)
  ANSI A58.1, 174–79
  ASCE Standard, 175–79
  Australian, 137–38, 183
  British, 182–83
  Canadian, 129, 138, 181–82
  deemed-to-comply, 188–90
  definition, 174
  European, 186–87
  international, 181–87
  ISO, 186
  Japanese, 185
  purpose, 174

Canadian National Building Code, 129, 138, 181–82
Canopies, wind effect on, 83
Canyon effect on wind, 60
Chimneys, vibration of, 113–14, 138–40
Chinook, 23
Cladding wind forces, 98
Cliff effect on wind, 61–62
Codes (*See* Building codes)
Columbia, Missouri:
  directional effect on high winds, 66
  extreme wind data, 26–27, 30
  microburst wind speed record, 21
Concrete-block buildings, 189–90
Connection failure, 142–44
Coriolis force, 2, 11, 13
Correlation coefficient, 53–54
Correlation of wind pressure fluctuations, 91
Correlation of wind speeds between sites, 31
Critical Reynolds number, 93, 111
Cyclones, 2 (*See* Hurricanes)
Cylinders:
  drag on, 99–100
  dynamic lift on, 114–15
  flow pattern around, 93
  Reynolds number effect on, 92–94
  wind pressure on, 92–96

Dampers:
  tuned-mass, 126
  viscoelastic, 126
Damping ratio, 134, 136
Darwin wind disaster, 142
Deemed-to-comply standards, 188–90
Dimensionless pressure (*See* Pressure coefficients)
Directional preference of high winds, 66–67
Dominant response frequency, 129

Index

Downbursts, 20–22
Drag, 98–100
  buildings, 98–99
  circular cylinder, 99–100
  fences, 100
  friction, 105
  rectangular cylinder, 99–100
  skin, 105
  surface, 45
  trussed towers, 101
Drag coefficient, 98–100 (*See* Drag)
Dust devil, 19
Dynamic pressure, 72
Dynamic response of buildings and structures, 109–26
  across-wind response, 136–40
  along-wind response, 131–34
  Australian Standard method, 137
  Canadian Standard method, 138–40
  damping ratio, 134, 136
  equivalent static-load approach, 127
  frequency-domain approach, 127
  general approaches to predict, 126–31
  natural frequency, 134–36
  Solari's method, 132–34
  time-domain approach, 127

Ekman spiral, 46
Escarpment effect on wind, 58
European Community Standard, 186–87
Exceedance probability, 25–27, 33, 36, 37–39
Exposure categories, 43–45
Extreme winds, 1–24 (*See* High winds)
  map of ANSI, 177

Fairing, 125
Fastest-mile wind, 26, 47
Fatigue:
  high-cycle, 140
  due to hurricane wind, 142
  on joints, 142–43
  Kemper Arena, damage to, 143
  low-cycle, 140
  low-rise metal buildings, 142
  metal, 141
  mitigation of, 144
  Palmgrem-Minor theory, 140
  roof damage due to, 142
  Tacoma Narrows Bridge destruction of, 144
  test result, typical, 141
  testing of roofs in Australia, 142
  wind-induced, 140–44
Fences:
  drag on, 100
  wind effect of, 83
Fisher-Tippett distribution, 30
Flow separation, 73, 77
Flutter, 122
  classical, 122
  panel, 123
  single-degree-freedom, 123
  stall, 122–23
Foehn, 23
Forces:
  cladding, 98
  drag, 89–101
  lift, 101–3
Form drag, 105
Free-stream turbulence, 91
F-scale, 17–18, 35
Fujita scale, 17–18, 35

Galloping, 115–19
  across-wind, 115–16
  criterion, 117
  Glauert-Hartog criterion, 117
  transmission lines, 119
  wake, 119–20
Glauert-Hartog criterion, 117
Gradient wind, 41
Gust effect factor, 49, 128
Gust factor, 49, 128
Gust response factor, 49, 128
Gust speed (gust velocity), 48

Helmholtz frequency, 89–90
Helmholtz oscillation, 89–90
High winds, 1–24
  amplification by hills and mountains, 56
  correlation between sites, 31
  cyclones, hurricanes and typhoons (*See* Hurricanes)
  different names, 1
  dust devils, 19
  mountain-downslope winds, 22–23
  probability, 25–39

High winds (*cont.*)
  thunderstorm or straight-line winds, 20–23
  tornadoes (*See* Tornadoes)
  types, 1
  water spout, 19
Hill effect on winds, 56–60
Hot-film anemometer, 170–71
Hot-wire anemometer, 170
Hurricanes, 2–10
  causes of damage by, 7
  direction of rotation, 2
  flow pattern, 3
  frequency, 9
  intensity ratings, 7 (*See also* Saffir/Simpson scale)
  life span, 2
  maximum speed, 10, 56
  model, 3 (*See also* Rankine Vortex model)
  movement, 2
  origin, 2
  pressure distribution, 3
  season, 9
  setup, 8–9
  shape, 2
  size, 2
  structure, 2
  surface wind, 5
  surges, 7–9
  translational speed, 2
  wind speed, 3, 10, 56
  wind tide, 8–9
  worst disasters, 7–8
Hyperbolic cooling towers:
  mean pressure coefficient, 96–97
  rms (root-mean-square) pressure coefficient, 97

Integral length scale, 54
Integral time scale, 54
Intensity of turbulence, 50
Interference, 123
Internal boundary layer, 44
Internal pressure, 70, 84–90
  blast effect, 89
  change during tornado passage, 16
  change with openings, 84–85
  coefficient, 84
  Helmholtz oscillation, 89–90
  leakage (crack) effect on, 88–89
  model test, 169–70
  predicting, 86–89
  safety implications, 86
  steady, 86–89
  unsteady, 89–90
International standards, 181
ISO Standard, 186

Japanese Standard, 185
Jet stream, 10

Kansas City, extreme wind probability of, 36
Karman constant, 42
Karman vortex street, 93
Kemper Arena wind damage, 21

Length scale:
  definition, 54
  free-stream turbulence, 91
  signature turbulence, 91
Lift, 101–3
  coefficient, 102, 114
  dynamic, 102–3, 114
  rms (root-mean-square), 103
  steady-state, 101
  unsteady, 102
Limit-states design, 184
Local mean wind speed, 47
Lock-in, 113
Logarithmic law, 42
Long-term average wind speed (*See* Annual mean wind speed)

Mach number:
  definition, 164
  effect on pressure coefficient, 164
Macrobursts, 21
Manometers, 171
Map of extreme winds in United States, 177
Masonry building, 189
Massachusetts Institute of Technology Earth Science Building, arcade wind effect on, 65
Mean velocity, 47
Melbourne University fatigue test of roofs, 142–43
Microbursts, 21
Mitigation of wind damage, ASCE Task Committee report, 188

Model codes, 173, 179–81
Moment by wind:
  horizontal (torsional or twisting), 104
  vertical (overturning), 103
Monte Carlo simulation of hurricanes, 34
Mountain downslope winds, 22–23, 56
Mountain effect on winds, 55–62

Natural frequency:
  buildings, 134–36
  cantilever beam, 135
  stacks, 135
  structures, 134–36
Normal forces by wind, 103
North Carolina Code, 188–89
Nuclear power plant, tornado design criteria of, 19–20, 37

Octagonal cylinders:
  drag and lift coefficients, 119
  galloping of, 119

Palmgrem-Minor theory, 140
Parapets:
  effect of, 83
Partial duration series, 32
Peak factor, 127–30
Peak wind speed, 47–48
Performance-type codes 187–90
Pitot tube, 70, 170
Power law, 42–43
Power-law exponent, 43
Power spectrum, 51–52
Prescriptive-type codes, 187–90
Pressure (See Wind pressure)
Pressure coefficients:
  area, 73
  buildings, 73–78
  definition, 72
  external, 72–74
  hyperbolic cooling tower, 96–97
  internal, 84–90
  local (local mean), 72
  mean, 73
  peak, 74
  roof, 81
  roof corner, 76
  rms (root-mean-square), 74–75
Pressure drag (form drag), 105
Pressure fluctuation correlation, 91
Pressure gust factor, 49
Pressure transducers, 171
Probability:
  density function, 28
  distribution function, 28
  exceedance (See Exceedance probability)
  extreme wind, 25–39
  hurricane wind speed, 32–34
  ordinary wind, 28–29
  thunderstorm wind, 30–31
  tornado strike, 34–35
  tornado wind speed, 34–37
Probability density function, 28
Probability distribution function, 28
  Fisher-Tippett, 30
  Gumbel, 30
  Rayleigh, 28
  Type-I, 30, 32
Prototype measurements, 70–71

Rankine vortex model, 3, 13
Rayleigh distribution, 28–29
Reattachment, 78–79
Rectangular buildings (See Buildings)
Rectangular cylinders:
  drag on, 99–100
  lift on, 117
Recurrence interval (See Return period)
Reference pressure, 70
Relative intensity of turbulence, 50
Resonance:
  cylinder, 112
  Helmholtz, 89–90
Response (See Dynamic response of buildings and structures)
Return period:
  definition, 26
  significance, 37
  Type-I, 30
Reynolds number:
  critical, 93, 111
  definition, 92
  effect on pressure coefficient, 164
  hypercritical, 94
  subcritical, 94
  supercritical, 94

Roof:
 damage, 142–43
 fatigue tests, 142–43
 flow around, 79–83
 freestanding, 81–83
 overhang, 83
 pressure coefficient, 81

Saffir/Simpson scale, 7
Santa Ana wind, 23
Scaling laws, 165
Sears Tower, ground-level gust around, 63
Separation zones, 73, 77–80
Setup (wind tide), 8–9
Shear:
 on buildings, 104
 on ground surface, 45
Shear velocity, 42
Shedding (See Vortex shedding)
Shielding effect of hills, 57, 61
Signature turbulence, 91
Similarity in wind tunnel tests, 162–64
 approaching-flow similarity, 163
 dynamic similarity, 162
 geometric similarity, 162
 kinematic similarity, 162
 Mach number, 164, 166
 Reynolds number, 164, 166
 scaling laws, 165
Skin drag, 105
Skyscraper-spawned winds, 63
Solari's method, 132–34
Solidity ratio, 101
South Florida Code, 189–90
Spectral density function, 51–52
Spectrum of wind turbulence, 51–52, 129
 peak, 52
 typical, 52
Speed-up ratio, 57
Spires, 159, 161
Spoilers, 126
Stagnation point, 71, 73
Stagnation pressure, 71
Stakes, 125
Stall, 102
Standards (See Building standards)
Stilts' effect on winds, 65
Storm surges, 7–9
Straight-line winds, 20–23

Strouhal number, 111
Suction vortices, 19
Surface drag coefficient, 43, 45
Surface wind, 5, 40

Tacoma Narrows Bridge, destruction of, 144
Taylor's hypothesis, 54
Temporal (time-averaged) mean, 47
Terrain roughness, 43
Terrain roughness effect on wind, 43, 45
Thunderstorm winds, 20–23
 probability of, 30
Time scale of turbulence, 54
Topographical effect on winds, 55–62
Tornadoes, 10–20
 atmospheric pressure change, 14–17
 building pressure drop, 15–17
 cause, 10
 Codell, Kansas, 10
 design criteria, 19–20, 37
 detection, 11
 direction of rotation, 11
 Fujita scale, 17–18
 geographical distribution, 10
 largest, 11
 life span, 11
 Los Angeles, 10
 in Midwest of United States, 10–11
 in Missouri, 12
 nuclear power plant design criteria for, 19–20, 37
 path, 11
 pressure field, 13
 probability of, 34
 rotational velocity component, 13
 seasonal variation, 12
 shape, 11
 size, 11
 strike probability, 34–35
 suction vortices, 19
 translational speed, 11
 Tri-State, 11
 velocity fields, 13
 wind-speed probability, 34–37
Tornado belt, 10
Tornado-like vortices, 19
Tornado season, 12

Torsional divergence, 120–22
Torsion on buildings, 104
Trees, effect on wind, 62
Turbulence:
  autocorrelation coefficient, 53
  characteristics of winds, 46
  correlation, 53
  eddy transport velocity, 54–55
  integral scales, 54
  intensity, 49–51
  level, 50
  relative intensity, 50
  spectrum, 51
  Taylor's hypothesis, 54
Type-I distribution, 30–32
Typhoons, 2 (*See* Hurricanes)

Urban street canyons, 64

Velocity gust factor, 49
Velocity pressure, 71, 72
Venturi effect, 60
Vibration (*See also* Wind-induced vibration):
  basic concepts, 193–98
  critical damping constant, 194
  cylinders, 198
  damped, 194–96
  damping ratio, 194
  frequency ratio, 196
  logarithmic decrement, 195
  magnification factor, 196–97
  mechanical admittance function, 196
  model tests 167–69
  modes, 130–31
  peak, 132
  undamped free, 193–94
Von Karman constant, 42
Vortex shedding:
  chimneys (stacks) 113–14
  circular cylinders, 93, 102
  dynamic lift generated by, 114–15
  non-circular cylinders, 114
Vortex-shedding vibration, 111
Vortex spoilers, 125–26

Wake buffeting, 123
Wake of building, 73, 77
Water spout, 19
Windbreaks, 62, 83
Wind characteristics, 41–68

Wind damage:
  aircraft, 22
  corrugated roofs, 142–43
  Darwin, Australia, 142
  Kemper Arena, 21
  masonry walls (unreinforced), 86
  roofs, 86, 142–43
  Tacoma Narrows Bridge, 144
Wind direction preference, 66–67
Wind direction variation with height, 46
Wind forces, 98–103 (*See also* Drag; Lift)
Wind-induced fatigue (*See* Fatigue)
Wind-induced vibration, 109–40 (*See also* Vibration; Dynamic response of buildings and structures)
  across-wind, 136–39
  along-wind, 131–36
  bridges, 121, 144
  buffeting, 83, 110, 123–24, 131
  buildings, 124, 132–34
  cantilever beams, 134
  chimneys, 113–14, 138–40
  at critical Reynolds number, 115
  dampers, 126
  damping ratio, 134, 136
  factors contributing to, 109–10
  flutter (*See* Flutter)
  galloping, 115–17
  lock-in, 113
  means to reduce, 124–26
  stacks, 113–14, 138–40
  torsional divergence, 120–22
  towers, 138–40
  transmission lines, 116, 119
  two-dimensional structures, 110
  types, 111–24
  vortex shedding (*See* Vortex shedding)
Wind moment, 103–4 (*See* Moment by wind)
Wind pressure, 69–97 (*See also* Pressure coefficients)
  absolute, 70
  ambient, 70
  circular cylinder, 92–96
  coefficient (*See* Pressure coefficients)
  cooling towers, 96–97
  correlation, 91
  definition, 69

Wind pressure (cont.)
dynamic, 72
external, 70, 72–74
gage, 70
hyperbolic cooling towers, 96–97
internal, 70, 84–90
prototype measurements, 70
rectangular buildings, 73–78
roof corner, 76
stagnation, 71
suction, 67
units, 69
velocity, 72
Wind-resistant construction, 188–90
Wind-sensitive structures, 109
Wind spectrum, 51–52
Wind speed:
amplification by buildings, 62–65
amplification by topography, 55–62
annual mean, 47
correlation between sites, 31
fastest-mile, 26, 47
gust, 48
highest measured, 56
hurricanes, 3
local mean, 40, 47
logarithmic-law, 42
map of the United States, 177
maximum, 56
power-law, 42
probability, 25–39
temporal (time-averaged) mean, 47
tornadoes, 17–18, 20
variation with height, 41–42
variation with surface roughness, 43
Wind tide (set up), 8–9
Wind tunnel models, 167 (See also Wind tunnel tests)
bridges, 150
equivalent, 167
replica, 167
section, 150, 169
Wind tunnels (See also Wind tunnel models; Wind tunnel tests):
boundary-layer, 78, 155, 159–61
for bridge testing, 158–59
closed-circuit, 151–52
commercial, 157–60
components, 152, 161–62
cross-sectional geometry, 156
drive sytems, 156
flow circuits, 151
guide vanes, 152, 162
high-speed, 155
hypersonic, 155
largest in the world, 153
low-speed, 155
meteorological, 156–57
NASA Ames Research Center, 153
open-circuit, 151–52
pressure condition, 154
pressure gradient, 160–61
pressurized, 154–55
screen, 162
special-purpose, 158
speed, 155
straighteners, 152, 161
supersonic, 155
taxonomy, 151
temperature stratification, 156
test section, 161
throat condition, 153
turbulence level, 156, 162
turntable, 161–62
types, 151
velocity profile, 155
water-wave, 158
wind speed, 155
Wind tunnel tests, 147–72 (See also Wind tunnels)
ambient pressure, 70
bridge models, 150
Cauchy number, 167
circumstances for, 148–50
common types, 150–51
cost, 148
data acquisition systems, 172
dispersion of pollutants, 151
domes, 150
elastic models, 150
equivalent models, 167
factors requiring, 148–50
flow pattern around buildings, 150
gravity effect, 169
instrumentation, 170–72
internal pressure, 169–70
pedestrian winds, 150
performance criteria of structures, 150
purpose, 147
replica models, 167

rigid models, 150
roofs, 150
section models, 169
similarity parameters (*See* Similarity in wind tunnel tests)
snow drift, 151
structures requiring, 147
tall buildings, 150

urban wind environment, 151
vibrating structures, 167–69
Wind velocity (*See* Wind speed)
Wood-frame buildings, 189
Woods effect on winds, 55
World Trade Center:
 use of dampers, 126
 vibration due to interference, 123